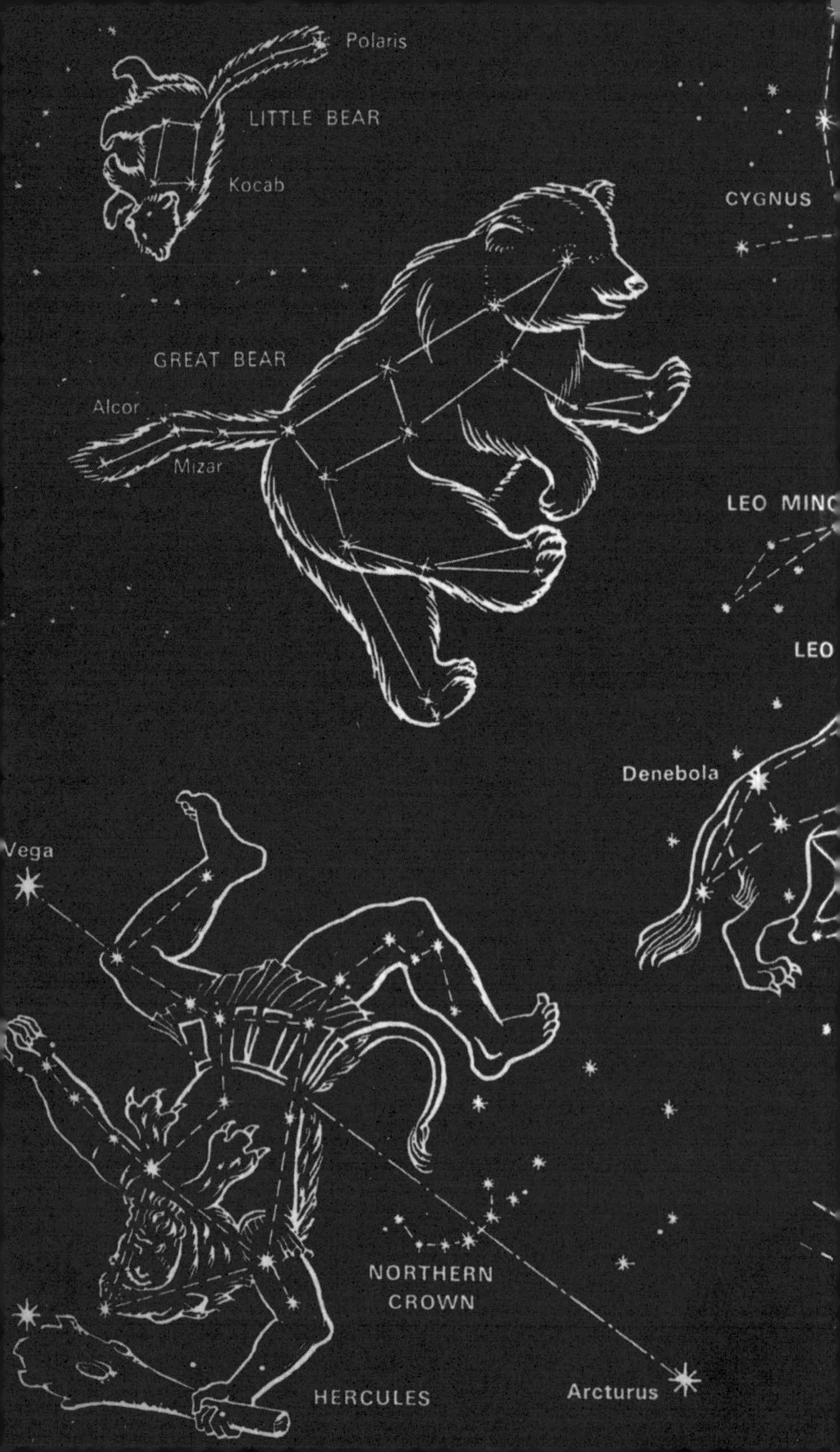

Polaris
LITTLE BEAR
Kocab
CYGNUS
GREAT BEAR
Alcor
Mizar
LEO MINO
LEO
Denebola
Vega
NORTHERN
CROWN
HERCULES
Arcturus

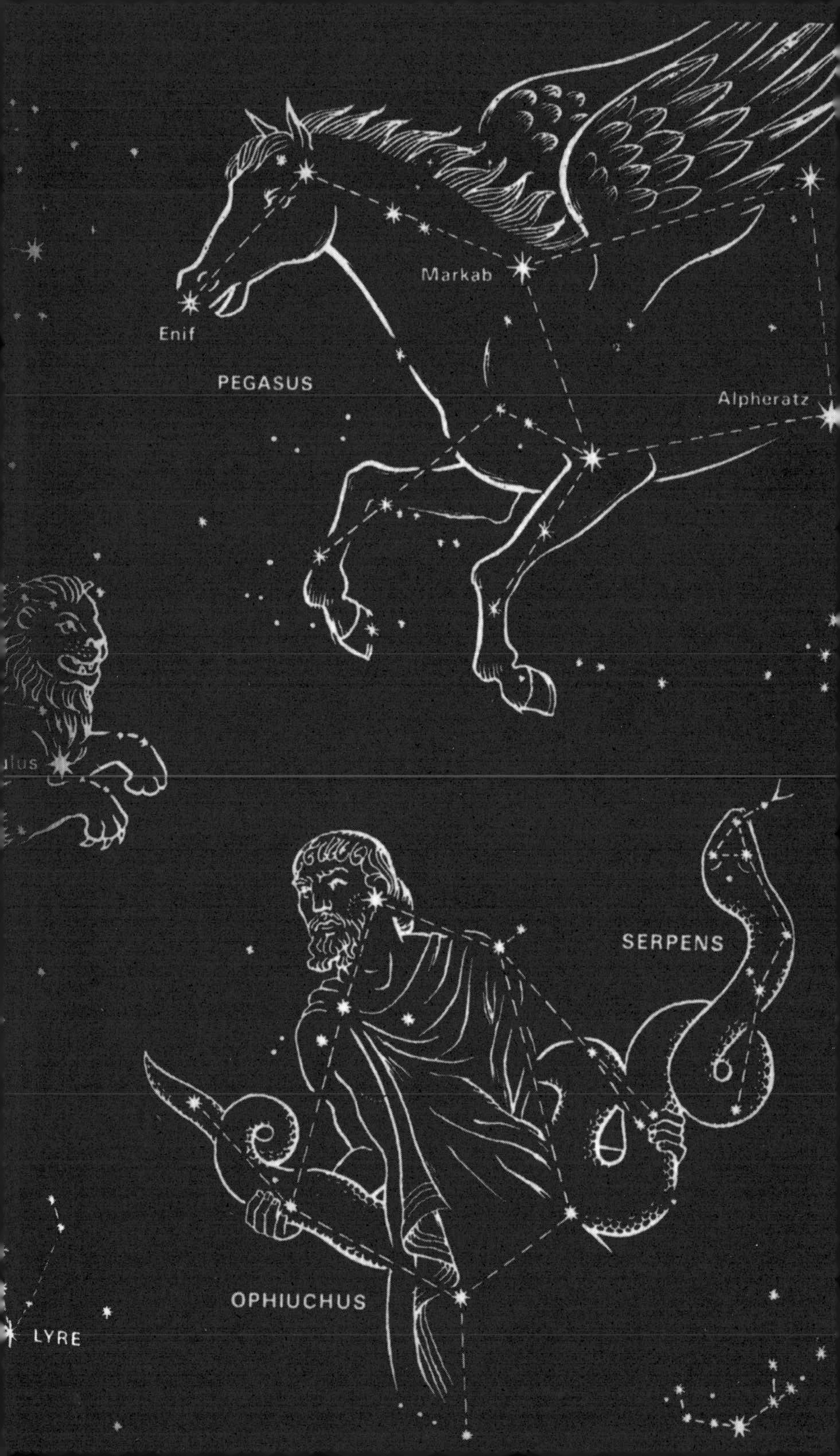

Markab
Enif
PEGASUS
Alpheratz
ulus
SERPENS
OPHIUCHUS
LYRE

PATRICK
MOORE

First published 1964
This edition published 2009

The History Press
The Mill, Brimscombe Port
Stroud, Gloucestershire, GL5 2QG
www.thehistorypress.co.uk

British Library Cataloguing in Publication Data.
A catalogue record for this book is available from the British Library.

ISBN 978 0 7524 4902 9

Typesetting and origination by The History Press
Printed in Great Britain

Contents

TO

ELEANOR R. DENNY

A Word from the Author

Do you like looking at pictures?

Probably you will reply 'yes', particularly if the pictures are well-drawn. Yet strangely enough, nearly all people over-look the most fascinating pictures of all, simply because they do not know where to look. The pictures are in the sky, formed by the patterns of stars, and you can see them on any clear night.

Let us admit that the pictures are not very clear. It takes a good deal of imagination to make a flying horse out of the star-group known to us as Pegasus, or a man's figure out of Orion. All the same, it is not hard to learn how to recognise these unusual pictures when some of the stories about them are known.

The people who first told these stories were the Greeks. Modern Greece is a small country of south-eastern Europe, with its capital at Athens; it is a fascinating place, and well worth a visit if you ever have the chance to go there, but is is not a great power in the same sense as Britain, America or Russia.

More than 2,000 years ago, things were very different. Ancient Greece was the leading country of Europe, and the Greeks were the most learned people in the whole world. They were scientists, writers, poets and athletes, as well as being brave and skilful fighters. Athens was a splendid city, and there were other Greek cities as well, such as Sparta and Thebes.

The Greeks were not Christians, and indeed there was no Christianity, since Jesus had not been born. Instead, the Greeks worshipped a large number of gods and goddesses, together with heroes and adventurers of all kinds. Each god had his (or her) own temples, and each was connected with some particular subject. For instance, Ares was the god of war, while Poseidon was the god of the sea.

Many stories were told, and were passed down first by word of mouth and then by writing. This was fortunate, since, although the glory of Greece has long since passed away, we still know the tales, and can enjoy them even though we know that they are only legends. We can do more: we can look upward on any dark night and see the pictures of the old gods and heroes, outlined by the groups of stars known to us as constellations.

The real trouble about the Greeks was that they could never agree among themselves. Two of the leading cities, Athens and Sparta, were drawn into a war which ended with the ruin of Athens and which weakened all Greece so much that things were never the same again. Then came the rise of Rome, and the gradual spread of the mighty Roman Empire, which stretched over much of Europe, and even reached as far as England. Greece was conquered, and did not become an independent country again until less than 200 years ago.

The Romans were quite unlike the Greeks. They had various gods of their own, but generally speaking they were much more interested in warfare than in the arts. The result was that they took over the Greek gods and heroes, and renamed them. Nowadays most people use the Roman names; thus the king of the gods, known to the Greeks as Zeus, has become Jupiter, while the Greek war-god Ares has had his name changed to Mars. So far as we are concerned at the moment, this does not really matter, and there is no reason why we should not use the better-known names, even though the gods themselves are Greek.

Finally, about 300 years after the birth of Jesus, the Romans became Christians, and the old gods were worshipped no more. It was not very long afterwards that the Roman Empire broke up, and the so-called Dark Ages began; but the legends were not forgotten, and never will be forgotten. They are just as interesting to us today as they were to the people of ancient times, who believed them to be true.

So let us go back 2,000 years and more, so that we can revisit those wonderful places in which the gods were all-powerful. There is much to tell, and I cannot hope to give you more than a few of the legends, but I will do my best.

P.M.

1

The Mighty Hunter

I cannot tell you just where the gods lived. Nobody really knows, but it was said that they dwelt in a palace in the sky, far above the mountain in Greece which we call Olympus. At any rate, they could look down upon the Earth, and they could make all manner of trouble for mankind if they chose to do so. Since many of the gods were bad-tempered, and were easily offended, it was always wise to be polite to them.

The king of the gods, and most powerful of them all, was Jupiter. He had not always been king; earlier, his father, Saturn, had ruled both sky and earth, but he had treated his children very badly, and it had been no real surprise when Jupiter had rebelled against him and taken the throne. Saturn was not killed, of course, since the gods were immortal, but he left Olympus and went to live on Earth, where he reigned happily for many years.

During the struggle, Jupiter had been helped by his two brothers, Neptune and Pluto. Since there could be only one king, the brothers had to be content with second best. Neptune took charge of the sea, while Pluto, who seems to have been a very dark, gloomy sort of person, preferred to reign over the Underworld, where the spirits of dead people went. It was a difficult task to approach the Underworld, and few people were anxious to go there until they were forced to do so. We can imagine Pluto sitting there on his throne, scowling around him and shutting himself away from all light and cheerfulness. Even his watchdog, Cerberus, was a terrifying creature with three heads, who never went to sleep and whose duty was to prevent any living man or woman from entering Pluto's realm.

Meanwhile, Jupiter must have had a much happier time in Olympus. He had a beautiful wife, Juno, and he had many companions, ranging from his fierce son, Mars, the god of war, to the cheerful, mischievous Mercury, who flew about upon winged sandals and who became the messenger of the Olympians. As for Neptune, he was perfectly content to stay in the sea, and liked being in the water much better than walking around on dry land. He was apt to grumble and complain that he should have as much power as his brother, but at least he could always amuse himself by causing tremendous storms and watching sailors being tossed about in their ships.

Like almost all the gods, Neptune had several children. One of them, Orion, was particularly strong, and soon made his reputation as a hunter. However, Orion's mother, Euryale, was not a goddess, and so Orion himself was not immortal. He was only a half-god, even though he had much more strength than any ordinary man.

'I fear nobody and nothing,' he used to boast. 'What animal can hope to injure me? Why, I am so tall that I can walk on the bottom of the sea without wetting my head; I can kill the fiercest lion with a single blow of my club, and I can bring down a stag by putting an arrow through its heart even when it is many miles away from me. If I chose, I could slaughter every animal on the face of the earth!'

'That is foolish talk,' his companions would say, looking uneasy in case the Earth-goddess should hear. 'Remember, you are not a god, and one day you will meet your match.'

But Orion would only laugh and race off to the hunt, accompanied by his favourite dog, Sirius. Fear was unknown to him; the more dangerous the chase, the better he liked it, and as time went by he became more and more sure that no living creature could hope to get the better of him.

We may suppose that Jupiter, looing down from his palace above Olympus, was not always happy about what was going on. Once, indeed, he made up his mind to interfere. Orion had been hunting in the mountains of Boeotia, a wild part of Greece, when he saw some strange figures among the trees, and dashed after them, shouting gaily and making ready with his arrows. For once he was mistaken; the figures were not those of animals, but girls – the Pleiades, the seven daughters of the giant Atlas. Orion soon saw that he had been tricked by the dim light, but he ran on, anxious to catch up with the frightened girls and tell them that he meant no harm. Not unnaturally, the Pleiades became even more terrified at the huge, wild-looking figure bounding after them, and in their fear they appealed to the gods. 'Save us!' they called. 'Save us, before we are caught!'

Jupiter heard, and decided to answer the call. At once he changed all the Pleiades into stars, snatching them up and placing them together in the sky. Orion stopped, amazed at the sudden disappearance of the girls. He looked around in vain; the Pleiades were nowhere to be found, and he never saw them again.

One day Orion sailed across the sea to Crete, a sunny island in the Mediterranean. There were wild animals in the woods there, just as in Greece, and he was anxious to hunt them. When he landed, he founded that he was not alone. Coming towards him was a beautiful woman, wearing a hunting dress, with her hair collected into a knot on her head and a deer's skin flung over her shoulders. Her legs were bare, and in her hand she carried a bow with a quiverful of arrows. Orion stared, and then realised that he was facing Diana, the goddess of the hunt. Diana's skill was just as great as his own, and she was a very powerful person indeed, since her father was none other than mighty Jupiter.

'What brings you here?' she asked in her deep, rich voice. 'Surely there are enough creatures for you in your own land. Or perhaps you have fought against a beast too strong for you?'

Orion threw back his head and laughed. 'That can never be. Goddess though you are, I will try my skill against yours. If you doubt me, let us go hunting together!'

You may be sure that Diana needed no persuading, and the two wandered all over Crete, doing their utmost to better each other. If Orion shot a deer at sixty yards, Diana would kill one at eighty; if the goddess brought down a wild boar at fifty yards, Orion would shoot one at a hundred. At last, when the hunt was over, Orion laid his bow aside, and spoke. 'Do you doubt me even now? My strength is at least equal to your magic arts. I am as quick as yourself, and no creature can escape me. Of all hunters, I am the mightiest.'

Diana said nothing. She had seen what Orion had not; out of the ground, close beside them, had crawled a creature which was as dangerous as any charging boar. It was a scorpion, full of deadly poison, and it was preparing to strike. We cannot know whether Diana had commanded it to appear; it may be that Juno, the queen of Olympus, had become angry at Orion's boasting and had decided to put an end to him. In a second it was all over. The scorpion had stung Orion on his bare heel, and the hunter fell, his club crashing to the ground and his great figure sprawling on the woodland.

And that might have been the end of the story, but fortunately the great doctor, Æsculapius, was not far away. He hastened to the spot, and knelt down by Orion's side. The giant was dead, but even death had no terrors for Æsculapius; in fact, Pluto had become afraid that he might make all men immortal, so that there would be no more spirits descending to the Underworld. It was then that Jupiter's voice was heard, rumbling and harsh. 'Wait, Æsculapius, I must make up my mind what to do.'

Æsculapius paused, and then Jupiter spoke again. 'Very well, then; bring him back to life, but he can no longer be allowed to roam the earth. I will place him in the sky, together with his dogs, so that he may shine there for ever.'

And so it was done. Orion, alive once more, was lifted up and set among the stars, where you may still see him to this day.

Orion competes with Diana

If you want to see Orion, look for him on a winter evening, high in the south. You may not recognise him at first, but you will soon make him out, because he is so brilliant. His shoulder is marked by the red star Betelgeux and his foot by the brilliant white Rigel, while his belt is shown by three bright stars in a line, and

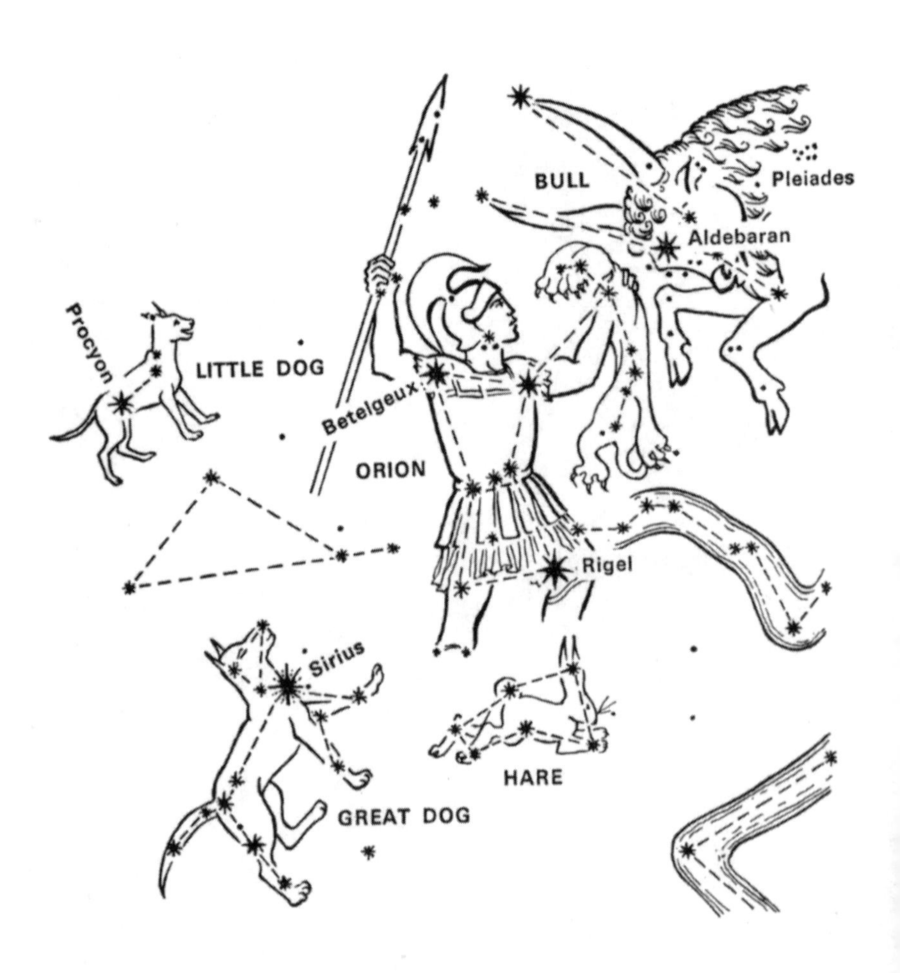

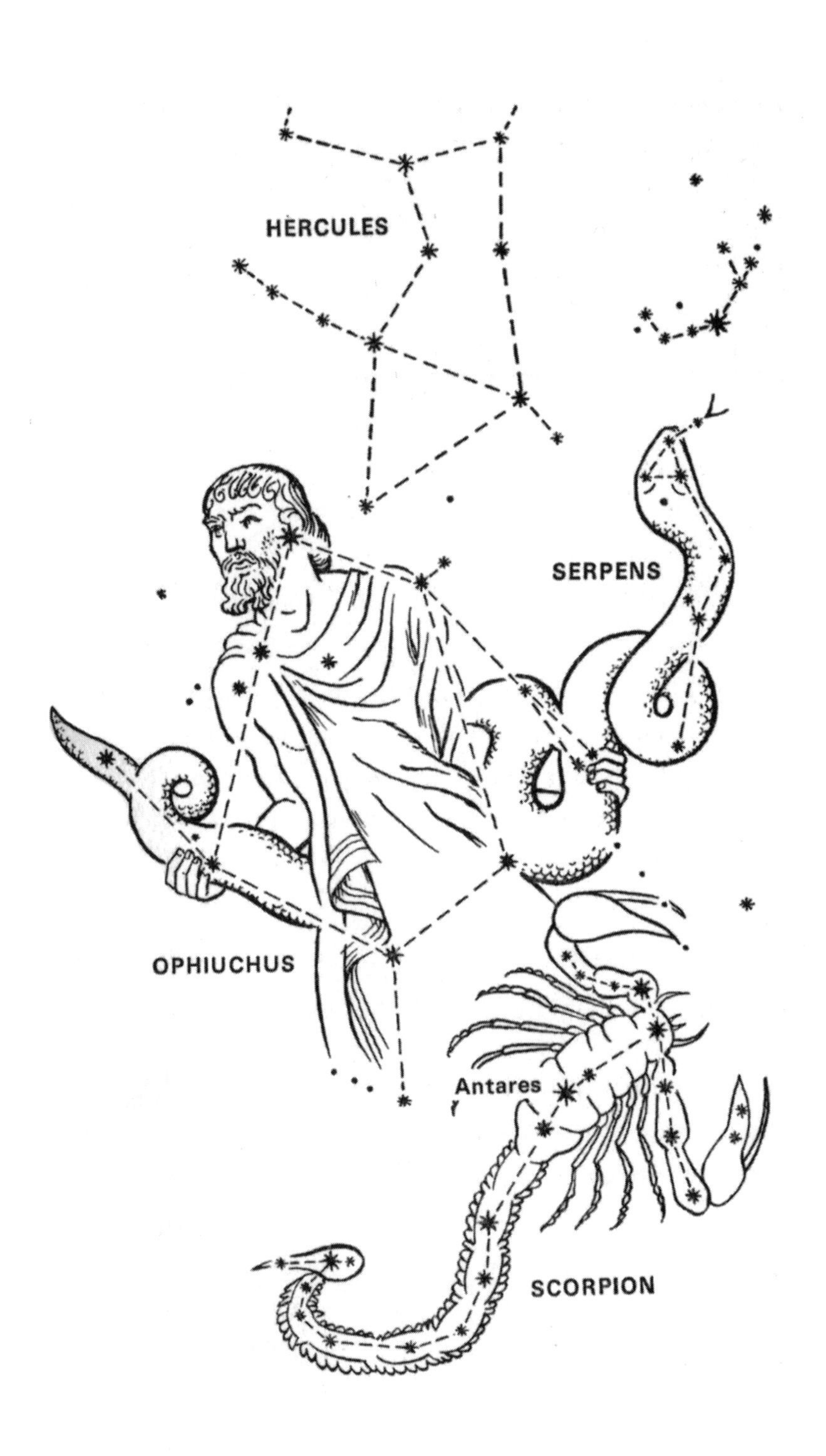
HERCULES
SERPENS
OPHIUCHUS
Antares
SCORPION

his sword by a misty patch known as the Great Nebula; astronomers have found that it is really a very large 'cloud' made up of dust and gas.

Orion's dogs are to be seen, too. His favourite, Sirius – the Dog-Star – lies in a line with the Belt, and is easy to find, since it is the brightest star in the entire sky. Near by are several other conspicuous stars, making up the outline of the Great Dog. The Little Dog has one bright star, Procyon, while below Orion lies the Hare, the animal which Orion is said to have been particularly fond of hunting.

The Pleiades are to be found in another constellation, the Bull. Follow the line of the Belt upward until you come to the bright red Aldebaran; then extend the line still further, curving it slightly, until you reach the Pleiades. All seven stars may be seen without a telescope on a clear, dark night, though in fact one of them is named after Atlas, the girls' father. At first sight the Pleiades look like a hazy patch, but if you look closely you will be able to make out the separate stars.

As for the Scorpion, it lies in the opposite part of the sky, so that it is above the horizon only when Orion has set and can do the Hunter no further harm. The Scorpion is a splendid constellation, but unfortunately it is always rather low in the sky as seen from Europe. On summer evenings you will be able to see the upper part of it, marked by one particularly bright and very red star known as Antares.

Æsculapius, the great healer, is represented by a much less conspicuous group, named Ophiuchus in the star-maps. Ophiuchus is situated above the Scorpion, and is best seen in the late spring and summer, but it is not really well marked.

Orion is probably the most glorious constellation in the whole sky. But this is natural enough; after all, he was once the mightiest hunter on the face of the earth.

2

The Gorgon's Head

In olden times Greece was not a unified country, as it is today. There were several kingdoms, and many of them were on bad terms with each other, so that wars were not at all unusual.

One of these kingdoms was Argos. Its ruler, Abas, was a great warrior, and he became so powerful that nobody even dreamed of disobeying him. Unfortunately he left not one son, but two. Prœtus and Acrisius were twins, and neither one would yield to the other. At first Prœtus was driven out, and made his way to King Iobates of Lycia (whom we shall meet again later).

'The land of Argos is mine,' said Prœtus proudly. 'My brother has no right to the throne. Give me an army, and I will be your friend for ever'.

King Iobates listened thoughtfully. He had no particular quarrel with Acrisius, but it would certainly be a good thing to have a friendly ruler next door to him; so he provided a strong force of men, and Prœtus led the way back into Argos. King Acrisius was ready for him, and there was a tremendous battle, in which so many soldiers were killed that both brothers saw that it would be senseless to go on fighting. They came to an agreement, by which they divided the land into two parts; Acrisius ruled one, and Prœtus the other.

King Acrisius had been badly frightened, and he knew that Prœtus would always be ready to attack him if possible. He needed a son to help him keep his country safe, but instead he had only a daughter, a beautiful girl named Danaë. So Acrisius did what he might have been expected to do; he went and consulted the Oracle.

When the gods of Olympus wished to talk to mortal men, they did so by means of an Oracle. The messages they sent were sometimes very muddled, but on this occasion Acrisius was given a straight answer to his question. No, he would never have a son; but he would have a grandson – and one day that grandson would kill him.

Acrisius was more frightened than ever, but for the moment there was nothing he could do except hope that Danaë would never have a child. Then, unhappily for herself, Danaë gave birth to a son. Acrisius looked at the baby boy, and was tempted to kill him there and then. He was a hard-hearted man, but even so he could not quite bring himself to kill his grandson at once. Instead, he shut both Danaë and the boy inside a wooden chest, and ordered it to be thrown into the sea, to drift wherever the ocean currents took it.

This was bad enough, and we can find no excuse for Acrisius, even though we may suppose that he felt ashamed of himself as he stood on the cliff and watched the wooden chest being carried farther and farther out to sea. At last he could glimpse it no more, and he turned back to his palace, feeling that he had made himself safe. It could not be long before the chest sank, taking Danaë and her boy to the bottom of the ocean.

What Acrisius did not know was that Jupiter, king of the gods, had been watching from above. Jupiter was able to control the sea – even though his brother Neptune was god of the waters – and he kept the ocean calm, so that instead of sinking the wooden chest drifted on. At last it was washed up on the beach of the island of Seriphos, where it was found by a fisherman named Dictys.

Dictys was surprised to see a wooden chest floating towards him, and even more surprised when he opened it and found a girl and a baby inside. Both were still alive, though they were exhausted and hungry, and Dictys at once took them to his house, giving them food and drink and promising to take care of them until they had fully recovered. 'What brings you here in this strange way?' he asked. 'Perhaps you have been shipwrecked? I am only a fisherman, but my brother Polydectes is king of this island, and if you wish I will take you to him.'

'I am a royal princess,' said Danaë proudly. 'My father is King Acrisius, of whom you have certainly heard. He must never know that I am alive, or he will send his men to hunt me down. Let us stay here, good Dictys, and live in peace.'

'That I will do,' said Dictys at once. 'My brother Polydectes must be told, but he is no friend of King Acrisius, and you will be quite safe. This is not a rich island, but at least there are none of your enemies here.'

So Danaë and her baby, who was named Perseus, lived quietly on Seriphos. As Perseus grew up, he became a tall, handsome boy who beat all his companions at running, and wrestling, and rowing; he was a musician too, and his good nature delighted all those whom he met. Acrisius, far away in Argos, had not the slightest idea that the grandson whom he had tried to kill had turned into one of the bravest and cleverest youths in all Greece.

Everything went well until Perseus was fifteen, but then he and Danaë were threatened with a new danger.

The trouble this time came from King Polydectes, who was by no means as good a man as his brother Dictys. Polydectes' wife had died, and the king wished to marry Danaë, who was a beautiful woman, as well as being of royal blood. Danaë did not want to become the king's wife, and when Polydectes asked her to marry him she refused at once. 'I can never marry again,' she pleaded. 'All my love is kept for Perseus, my boy. Do not keep on asking me, I beg you; my answer will always be the same.'

Polydectes pretended to agree, but there and then he made up his mind to get rid of Perseus. Without a son, Danaë might well change her decision; besides, the king did not dare to carry her off by force so long as Perseus was nearby.

Polydectes thought long and hard, until at last he hit upon a plan. He told his subjects that, since Danaë would not become his queen, he meant to take another princess as a bride, and he asked all his people to bring him a gift so that he could offer the princess some handsome presents. 'I am not a wealthy man, even though I am your king,' he said, 'and I know that you will help me. One horse from each of you – that is all I ask.'

Polydectes then gave a great banquet, to which all his subjects were invited. Dictys and Danaë were there, of course, and so was Perseus, who stood out at once because of his tall figure, even though he was wearing rough sailor's dress. One by one the people brought Polydectes their presents, but when it came to Perseus's turn there was a sudden silence.

'I have no present for you, my lord,' said Perseus simply. 'I would have brought one, but I have no money or land.'

Polydectes scowled. 'This is a fine way to behave,' he said angrily. 'You have been made welcome here, and you have lived on my island since you were cast up on the shore many years ago. You show little gratitude, young man.'

'I am not ungrateful,' said Perseus, looking at Dictys uncomfortably. 'I would bring you any gift in the world, if it were within my power to do so.'

Polydectes paused, and a smile came to his thin lips. 'Do not be so boastful. I can think of many gifts which you would certainly not want to bring to me, and I fear that you are untruthful as well as ungrateful.'

Perseus drew himself up. 'That is a lie, my lord, and I will obey any command that you care to give me.'

Polydectes stood up on the steps of his throne, and thrust his head forward. 'Very well, young man, I will take you at your word. Go, then, and bring me back the head of Medusa, the Gorgon!'

At once there was a murmur from the crowd, and Dictys turned pale, while Danaë let out a cry of despair. All of them knew that Polydectes was asking for the most dangerous gift in the whole world. Medusa was one of the three sisters who lived in a far-off land; the Gorgons, as they were called, had the bodies of women but snakes instead of hair, while their glances were so terrible that a single look at them was enough to turn any man into a block of stone. Even to come within sight of the Gorgons meant certain death, and to creep up on them unseen was almost impossible. Perseus knew this quite well, but he was too brave and too proud to draw back.

'I accept your challenge, my lord,' he said in a bold, loud voice. 'Give me a little time, and I will be back, bringing with me the Gorgon's head. I will show you that I am neither a boaster nor a coward.'

Polydectes laughed, and so did his courtiers and friends, but Perseus did not care. He turned and strode out of the great hall, the king's laughter ringing in his ears. He was not afraid; he had given his promise, and he meant to keep it.

Yet where could he start? The Gorgons were far away, and to reach their land would be difficult enough even if he knew which route to take. 'It will mean going on a ship,' thought

Perseus to himself, 'but that means I must have a crew, and who will dream of coming with me on an errand such as this?' Deep in thought, he made his way up to the top of the cliff which overlooked the sea-coast, and he stared out across the water towards the setting sun.

Suddenly he gave a cry of surprise. A tall, beautiful woman was standing close beside him, dressed in long blue robes and carrying a shield and a spear. Her expression was serious, and yet Perseus knew at once that she was to be his friend. There, too, was a young man hardly older than Perseus himself, who looked thorough cheerful and mischievous, and who seemed ready to burst out laughing at any moment. In his hand was a long staff, and Perseus could see that there were live snakes coiled around it.

'Who are you?' he asked fearlessly. 'I think you are not mortals, and that you come from Olympus.'

'Of course we're gods, which is lucky for you,' said the young man cheerfully. 'You should recognise Minerva, surely? Don't be put off by her solemn look. After all, she is the goddess of wisdom, so she can't go around laughing and chuckling all the time.'

Minerva smiled, and looked even lovelier than before. 'Mercury, your tongue will cause you trouble one day, even though you are the messenger of the gods and welcome wherever you go. Be silent, and listen to what I have to say.' She turned to Perseus. 'You have undertaken a dangerous task, foolish boy. Do you really think that you will be able to reach the Gorgons' land and then cut off the head of Medusa before she can turn you to stone? Many men have tried; all have perished.'

Perseus looked troubled. 'I do not know. At least I can try, and it is better to die bravely rather than live as a coward.'

'Well spoken,' said Mercury approvingly, and gave a playful jump into the air. He seemed to rise to a great height before he came down, and Perseus could see that there was something strange about the sandals he was wearing: wings sprouted from them, so that this remarkable young man could fly as easily as any bird.

'You have courage, and you will need it. If you fail, nothing can save you from being turned into stone,' said Minerva, and took off her shield. 'I will do what I can for you, but you must obey my orders without question. Take this shield; it is not heavy, but you will see that it is polished, so that it gives a picture in the same way as a mirror. When you come up to Medusa, do not look at her, but reflect her face into the shield. If you can do this, you will be able to strike her down before she can turn her eyes towards you.'

Perseus took the shield, and put it over his arm; it felt light enough, but it gave him a great feeling of strength. 'I have no sword, but I have a knife. Once I come within reach of Medusa, I know I shall be able to kill her.'

'With a knife such as that?' broke in Mercury, scornfully. 'You would be wasting your time. Remember, the Gorgons have tusks like those of a wild boar; their hands are of bronze, and they have golden wings which allow them to fly almost as fast as I can.

Even if you cut off Medusa's head at one blow, her two sisters would make short work of you – and you would not be able to kill them also, because they are immortal. Feel this.' He passed across a long, tapering sword. 'That's a great deal better than your knife, my boy. Take care of it.'

Perseus stared, and felt the blade. 'With a shield and a sword from Olympus, I cannot fail.'

'You can fail very easily, if you lack courage. If you hesitate even for a second, you are lost,' said Minerva in her grave voice. 'Mercury has brought you a pair of winged sandals, so that you will be able to swoop down on the Gorgons from above, but there is one more thing that you will need – and which we cannot give you. You must have the helmet which belongs to Pluto, king of the Underworld. When you wear this helmet, you will be invisible, and even the Gorgons will not be able to see you.'

Perseus fastened the sandals on to his feet, kicking off his own rough, heavy shoes into the grass. 'You mean that I must first go to the Underworld?'

Mercury burst out laughing, and Perseus wanted to laugh too. 'Upon my word! You know very little about Pluto. He's a grim old fellow, and once he caught you – well, there you would stay. No, you must keep well away from the Underworld.'

'But if the helmet is there …' began Perseus.

'Don't talk so much. Pluto's helmet is not kept in his kingdom of darkness; if it were, it would be beyond your reach. Luckily, it is kept in the care of some maidens called the Hesperides, the daughters of the Evening Star, who live in an enchanted garden far away to the south. Ask them, and they will lend you the helmet until your task has been done.'

'You are sure of that?' said Perseus, rather doubtfully.

'They will lend you the helmet once you can find their magic garden,' said Minerva quietly. 'I cannot tell you the way there; first you must go north, to the frozen land which has no name, and find the Three Grey Sisters, who live by themselves and have only one eye and one tooth to share between them. Go, then – and do not lose your courage.'

She pointed to the edge of the cliff, and Perseus stepped forward, looking down at the wave-crests so far below. For a moment he paused, and once again Mercury gave his boyish laugh. 'Are you frightened already? Jump over the cliff, and you'll soon see how easy it is.'

Perseus took one more look at the waves. Then he held his breath, and jumped into space. He half expected to fall headlong, dashing himself to pieces on the rocks, but then he found that he was flying; the winged sandals served him well, and almost before he had recovered himself he was far out across the water, soaring along like an eagle. When he turned his head, he saw that the coast of Seriphos had faded into the distance, and that Mercury and Minerva were no longer to be seen.

He flew on and on, until Greece was far behind him, and the rich green plains had given way to a cold, unfriendly landscape. His thoughts went back to the court of King Polydectes, and he wondered what was happening to his mother

Danaë and his faithful friend Dictys; but he knew that he had many perils to face before he saw them again, if indeed he ever did so.

Still he kept northward, and after seven days and seven nights he came at last to the nameless land of the Three Grey Sisters. He found them easily enough, sitting by the shore, looking as ugly and evil as can be imagined. They chattered to each other, taking it in turns to use their single eye and single tooth.

Perseus landed beside them, and bowed politely. 'I wish you well, good sisters. You are old, and you are wise. Tell me, if you please, where I may find the daughters of the Evening Star.'

'What? What? Who can this be?' screeched one of the Sisters. 'Give me the eye, so that I may look at him.' She fumbled for the eye, and blinked. 'A young man from the outer world, if I am not mistaken. Who are you, and what are you doing here?'

Perseus bowed again. 'I am the son of Danaë, princess of Greece. I am here on an errand for my lord King Polydectes, who has commanded me to see the Gorgon Medusa and cut off her head. Tell me, please, where to find the Hesperides, who can lend me the helmet which will make me invisible.'

'Certainly not!' croaked the second Grey Sister. 'Go away, you impudent boy, and leave us in peace. Give me the eye, so that I may look at him for myself.'

Perseus crept forward, and waited his chance. As the Sisters passed the eye from one to the other, he leaped across, snatched the eye, and then rose into the air, hovering above the icy sea. 'I have said that I mean you no harm, but I must know where to find the daughters of the Evening Star. Unless you tell me, I swear that I will throw your eye into the waters, so that you will be blind for ever. Speak, before it is too late!'

You can well picture how the Sisters screeched and cursed among themselves, but they knew that there was nothing to be done, so at last they told Perseus the way. 'Fly south,' they said, 'straight towards the noonday sun. You will come to a great mountain, where the giant Atlas kneels, holding the sky and the earth apart. There you will find the daughters of the Evening Star, and if you are lucky you may be able to borrow the enchanted helmet. Now give us back our eye, and be off.'

Perseus handed back the eye, and thanked them politely, but the Sisters did not listen; they were too busy grumbling and muttering. As Perseus flew away he could still hear their screeching voices, and it was some time before everything was silent once more.

Back he went, out of the dark and nameless land, until he reached the great mountain which rose up towards the clouds. There he found Atlas, groaning under the weight of the sky, and he also found the beautiful Hesperides, who spent their days dancing and singing round a tree which was heavy with golden fruit, and guarded by a fierce dragon which never slept. Perseus folded the wings on his sandals, and landed close beside the giant, who looked at him with surprise.

'Who is this who wears winged shoes?' asked Atlas in a deep rumbling voice. 'Have you been sent by Mercury, or have you perhaps stolen his sandals?'

'These sandals were loaned to my by Mercury himself,' said Perseus proudly. 'The shield belongs to Minerva, goddess of wisdom. I have sworn to kill the Gorgon, but first I must have the magic helmet which will make me invisible. Tell me, please, where it may be found.'

'Kill the Gorgon, eh?' growled Atlas. 'Well, I have no love for Medusa, so I'm ready to help you. My nieces have the helmet in their care, so you can borrow it for a while – but be sure to bring it back safely, or you'll have King Pluto to reckon with.'

'I promise, by all the gods, I will bring the helmet back as soon as I have cut off Medusa's head,' said Perseus simply 'Must I go to the entrance to the Underworld, or is the helmet here in your magic garden?'

'Here, of course,' said Atlas. 'My nieces have no wish to go down into that gloomy place – and as for me, I'm too busy holding up the sky. Take it, then, and be as quick as you can.'

The Hesperides were clustering around, anxious to see this brave young man who was so ready to face the terrible Gorgon. They pleaded with Perseus to stay awhile, but although he would have liked to have lingered in that lovely garden he was impatient to be gone. As soon as he thrust the helmet on to his head, he vanished from sight. 'Which way shall I go, good giant Atlas?'

'Far across the horizon, until you come to the island where the Gorgons live,' said Atlas, and pointed. 'It will take you many days, and you will see no living thing; any creature who goes near is sure to be turned into stone. I very much fear that the same fate will befall you.' He shook his head sadly. 'You have a magic sword, an enchanted shield and a helmet of invisibility, but you will need them all.'

'I can try, and if I do not succeed I shall at least have done my best. Farewell, and my deepest thanks to you. When I return, I shall have Medusa's head with me.'

Again Perseus soared into the air, and flew off in the direction that Atlas had shown him. Before long the tall mountain was out of sight, and the friendly, grassy plains had faded from view; at last he came to a wild, lifeless land, where nothing stirred and nothing grew. Perseus shuddered as he looked down. Below him were rocks, stones and boulders, but somehow this place was even more dreadful than the icy country where he had found the Three Grey Sisters.

As Atlas had said, the journey was a long one, but at last he came in sight of an island. Perseus paused, listening for any sound that might mean danger. Then he heard a strange rustling, and he saw that on the shore of the island something was moving and glinting. Slowly he flew downward, and then stopped once more. Three figures lay by the edge of the sea, while all round there were rocks of curious shape. Perseus' keen eyes told him that some of these rocks, at least, looked like men and animals.

'What can they be?' he thought. Then he realised the truth; the rocks had once been living creatures, who had looked at Medusa and had been turned into blocks of stone.

We may suppose that even Perseus felt afraid. One glance at the Gorgon would be enough, and he might so easily become just another lump of rock, to stay for ever on this dreadful island. For a moment he was tempted to turn back, but then he remembered his promise to King Polydectes, and he forced himself to fly on. Now all three Gorgons were in view, sound asleep, their wings rustling and their bronze hands catching the light of the pale sun.

At once Perseus turned away, and held up the polished shield until he could see the figures of the Gorgons reflected in it. For a moment the middle figure came into view, and Perseus gave a gasp of horror. It was Medusa, without a doubt; there was the woman's face, and there were the snakes which made up her hair, hissing and spitting as though they hated everybody and everything in the world (as they doubtless did). Thanks to Pluto's helmet, the Gorgons could not see Perseus, but if Medusa awoke there would be no chance to strike.

Now Perseus was only a few yards away from the three sleeping monsters, moving slowly, so as not to make any noise with the wings of his sandals. His heart beat furiously, and again he felt afraid, but he knew that he must lose no time. He would be able to strike only one blow; if he failed, he would be lost. He pulled out the magic sword and edged forward, holding the shield in front of his eyes.

Then he raised the sword, and brought it down. His aim was true; the blow caught Medusa full on the neck, and as Perseus swooped down he was able to grasp the snaky hair. As he flashed upward again, Medusa's head was safe in his grasp, and he thrust it into his goatskin bag so that he should run no risk of looking at it. But what a roar the other two Gorgons set up! Screeching and howling, they flapped their wings and circled round, staring at Medusa's dead body, and sniffing to see whether they could catch the scent of their invisible enemy. Suddenly they wheeled round, and Perseus knew that, even though they could not see him, they could tell just where he was.

Perseus darted away, but for an instant he thought that he must be caught; the bronze claws of the Gorgons almost brushed him, and the screaming made his blood run cold. Then he was away, with the two hideous creatures in pursuit. 'Winged sandals, fly faster than you have ever done before!' cried Perseus, his voice almost drowned by the Gorgons' howls. 'Make speed, or there is no hope for me!'

He dared not look back, and it seemed a very long time before he could tell that the flapping of the creatures' wings was becoming fainter. At last he risked a glance over his shoulder, and saw that, fast though the Gorgons were, the sandals which Mercury had given him were even faster. Steadily he drew away, until at last the Gorgons had been left far behind, and all was quiet once more. Even so, it was not until Perseus had passed across the deserted land, and had come back within sight of the mountain where Atlas knelt, that he felt really safe.

He found the giant still on top of the peak, and still groaning under the weight of the sky which he was forced to hold up. 'So you have returned,' said Atlas in his deep voice, as Perseus took off the enchanted helmet and stood before him. 'Well, young man, did you manage to cut off the Gorgon's head?'

Perseus held out the goatskin bag. 'Indeed I did. I have kept my promise; here is King Pluto's helmet, and all that I need to do now is to return the sandals to Mercury and the shield to the goddess Minerva. After that, I can return to Seriphos and make my gift to King Polydectes.'

'There is one more task for you to do first,' said Atlas. 'It is a small one, but I ask you to repay your debt to me. Show me the head of Medusa.'

Perseus jumped back in alarm. 'I have the head covered in this goatskin – surely that is enough? If you take one glance at that terrible face, you will be turned into stone! That surely cannot be what you want?'

'That is what I want,' said Atlas, and gave a tired smile. 'For countless years I have knelt here, holding up the sky which would otherwise fall upon the earth. I can bear the weight no longer, but if I am turned into a rock I shall be able to carry the heavens for all time, free of pain. Show me the head.'

Rather sadly, Perseus drew out Medusa's head from under the goatskin, and turned it towards the giant, taking care not to look at it himself. At once Atlas became hard and grey and rocky; he was no longer a giant, but a huge rock upon which the sky could rest for ever.

It was with a feeling of sorrow that Perseus returned the helmet to the daughters of the Evening Star, and took his last look at the rock which had once been Atlas. Yet he knew that the giant's sufferings were over, and that he had been turned into stone by his own wish. Then Perseus took to the air once more, and flew back towards his home in far-off Greece.

It was a long journey, and it took him many days, even though the winged sandals carried him through the air more swiftly than any golden eagle. In time he came to a coast he knew, but he could see that there was something strange about it. The country below, which should have been green and pleasant, was ruined; here and there he could make out broken houses, but nowhere could he see the farms and towns which certainly ought to have been there.

'This is the work of some evil force,' thought Perseus, and he came lower, searching to see whether he could catch sight of any dragon or other unpleasant beast. 'It may be that there is a task for me to do yet before I take my gift to King Polydectes.'

Presently he saw what seemed to be a human figure, and he wished that he still wore the enchanted helmet; if he were faced with any monster, he would be in full view – though it seemed difficult to believe that any new monster could be so fearful as the Gorgons. But the figure below was no monster. It was a beautiful girl, and she was chained to a rock; her face was pale, and she looked around her with dread, as though expecting some demon to rise up from the sea.

As she caught sight of Perseus she gave a cry, and shrank back, covering her eyes as though thinking that he would run her through with his sword.

'Who are you, and why are you chained?' called Perseus, as he sank down through the air.'I am Perseus, son of the Princess Danaë, and I mean you no harm. It will take me only a few moments to break these chains–'

'I – I am Andromeda,' came the reply, though the girl was sobbing so much that it was difficult to understand what she said. 'My father Cepheus is king of this land. I am here to be sacrificed to a monster from the sea. Keep away, or you will be killed too!'

Perseus alighted on the sand, and drew his sword.'I do not fear the monster, whatever or whoever it may be. Wait, while I break these chains.'

Andromeda screamed.'I tell you that you must go – and quickly! You do not understand. My mother, Queen Cassiopeia, boasted that I was more beautiful than the Nereids, the nymphs who live in the sea. That is why I am here. Leave me, before it is too late.'

But Perseus was more determined than ever not to go until he had rescued the princess, who was certainly very beautiful indeed, and with whom he had fallen in love at first sight.'I assure you that I am afraid of no monster. How can it be that your father, the king, has left you here alone? He must be a very hard-hearted monarch.'

'I cannot blame him,' said Andromeda, between her sobs.'Listen. The Nereids are the daughters of Neptune, god of the sea, and they are said to be more lovely than any mortal girls. When Neptune heard of my mother's boast, he was so angry that he sent a monster to lay waste our land, and he will be satisfied only when I am dead. My father went to the Oracle, and was told that the only way to save his people was to chain me here to make a meal for the sea-monster. Now do you understand why you must leave me?'

'I do not,' said Perseus grimly, and felt in his goatskin to make sure that Medusa's head was still there. 'Sea-god or no sea-god, I will never leave you. Wait, and we will see what I can do with this fearsome creature that Neptune has sent to eat you.'

Andromeda could say no more. To defy one of the Olympians was a brave act indeed, particularly as Neptune was no less a person than Jupiter's brother, but by now the princess was so terrified that she was unable to believe that anybody could help her. For hours the two waited, crouched down by the rock, until the sun was setting across the water, and the light had faded to a golden yellow. All was quiet and still, but at last Perseus fancied that he saw a ripple on the glassy water, and he made ready.

'It is coming,' he said softly. 'Be brave, and put your trust in me.'

Slowly a monstrous head rose out of the water, and Perseus could see the gleam of long white teeth. The sea-beast crept towards him, snarling and hissing, until the whole length of its snake-like body could be seen; it was hundreds of feet long, and it seemed to breathe fire. Perseus' hand went to his sword, but then he drew back. There was a better way, but he must wait until the moster was close to him.

The snarling filled the air, and Andromeda shut her eyes, almost fainting with terror. Then, just as the monster prepared to spring, Perseus drew out the Gorgon's head, and held it up. There was a sudden silence; the snarling died away, and where the sea-creature had been there was nothing but a long grey rock.

You can imagine Perseus's relief as he hacked away Andromeda's chains, and you can imagine, too, the relief of King Cepheus and Queen Cassiopeia when their daughter was carried back to them unharmed. Cassiopeia wept loudly, and Cepheus was just as overcome. 'I cannot believe that we have been saved,' he said over and over again. 'My daughter alive, and Neptune's monster dead … It is too much! Stay with us, and live in my land. Take Andromeda in marriage, and rule over half my kingdom. We owe everything to you, everything, and you may have whatever you desire.'

Perseus looked at Andromeda, and the princess smiled back at him. 'I am grateful, my lord. There is nothing I want more than to marry your daughter, and I will do my best to make her

Perseus saves Andromeda from the monster

happy, but I cannot stay with you; my home is far away, and although I have killed the Gorgon I have not yet kept my promise and given her head to King Polydectes. You will understand, I know.'

'This is a pleasant land,' said Cepheus sadly. 'Take all of it, if you wish. I would count it an honour to become one of your subjects.'

Perseus smiled again, and shook his head. 'I am not asking for honour or wealth, my lord. I cannot leave my own people, but I will guard your daughter with my life, and I will bring her to see you often.'

Cepheus knew that he could not make Perseus change his mind, but at least he persuaded him to stay for a while so that the wedding could be celebrated. And a glorious wedding it was; all the people of the country came, and the feasting and dancing lasted for seven days. Then, just before the celebrations were over, King Cepheus drew Perseus aside and asked him a question. 'You have told me that you were given winged sandals by Mercury, and an enchanged shield by Minerva. What do you mean to do? Keep them, or give them back?'

Perseus drew himself up. 'To keep them would be unworthy as well as ungrateful. I will return the sandals and the shield as soon as I am back in Seriphos; the gods will know where I am, and they will come to me. As for the Gorgon's head, I must give it to King Polydectes, as I promised to do, but I do not think that he will wish to keep it for long.'

'That is the answer I expected from you,' said Cepheus, 'and it makes me even more sad that you should have to leave us. Go, then, but do not forget that you have made a promise to come back.'

Perseus swore that he would not forget, but he too felt sad as he sailed away with his princess and watched Cepheus and Cassiopeia waving to them from the shore until they were out of sight. He still had the winged sandals, but he travelled by ship, rowed by a crew of the strongest and most trustworthy men in Cepheus' kingdom. He could hardly carry Andromeda in his arms the whole journey, and besides, there were the many presents that Cepheus had made him accept.

The ship seemed very slow compared with the winged sandals, but at last they came within view of the coast of Seriphos, and Perseus pointed. 'That is the land where I have lived for almost all my life, though whether we shall live there in the future I cannot tell.'

'It looks cold and rocky,' said Andromeda, and drew close to him. 'I am not sure that King Polydectes will be glad to see you. From what you have told me, he is not a warm-hearted man.'

Perseus said nothing, but as soon as the ship touched the coast he went ashore, leaving his princess in the care of the sailors. He made his way towards the city, and paused as he came in sight of the royal palace. Then there was a cry, and he saw his mother running towards him.

'Mother!' cried Perseus, and put his arms round her. 'You must have known that I would be back. I have kept my promise, and nobody can hurt you now.'

Danaë was too overcome to speak, but presently she recovered. 'I could not dare to hope that you would return,' she said, the tears streaming down her face. 'Polydectes believes you to be dead, and he has told me that I must become either his wife or – or else his slave. The good Dictys is in danger, too. Even now the king is in his palace, feasting with his courtiers and making arrangements for our wedding. I cannot marry him, Perseus – I will not!'

'You shall not,' promised Perseus, and looked stern. 'I have a surprise for Polydectes. Stay here, Mother; do not come near the palace.'

In he strode, looking very different from the boy he had been when he had set out. As he came into the banqueting hall, there was a sudden hush. Polydectes, seated on his golden throne, turned pale and stammered. 'It is you, young man. So you could not dare to face the Gorgon, after all?'

'I have carried out the task you set me,' said Perseus, his voice ringing out across the great hall. 'You thought that you had sent me to my death, but you have failed – and now you must pay the penalty. You asked for Medusa's head, Polydectes. Well, here it is!'

He threw back the goatskin, and swung the head towards the royal throne. Though Medusa was dead, the snakes of her hair were still alive, and filled the air with their hissing and spitting. But Polydectes did not hear; he could no longer hear anything. Like the sea-monster, he had become a grey rock, and his evil life had come to an end.

It would take me too long to tell you how Perseus settled the affairs of the country, and set up his old friend Dictys as king in Polydectes' place. But he still wanted to see his old home, and so after a while he set out for Argos, taking Danaë with him as well as Andromeda. He had no thought of taking his revenge on King Acrisius; by now his grandfather was a very old man, and Perseus felt that he would, after all, be glad to know that he had not been responsible for killing either Danaë or her baby.

Yet the story had an ending which was, in its way, sad. Perseus did not tell the king who he was; indeed, Acrisius was busy attending some games which were being held in honour of a visiting ruler. There were races, and wrestling competitions, and throwing the javelin; Perseus took part, and carried off the prizes in all the games for which he entered.

At long last there came the discus-throw, and Perseus braced himself for a mighty effort. He could see King Acrisius on his throne, his white beard flapping in the breeze, and Perseus said to himself: 'I will win this throw, and then I will go up to my grandfather and tell him who I really am.'

He drew back, and hurled the discus – a heavy wooden ring which could soar for a great distance. At that moment a sudden gust of wind came in from across the sea, and swung the discuss to one side. By ill-fortune it fell upon King Acrisius, and as Perseus leaped towards the throne he could see that his grandfather was dead.

The Oracle's prophecy had come true, but not in the way that Acrisius had expected. Though Perseus was deeply grieved, there was nothing he could do except tell the crowd the whole story of what had happened. Argos had lost its king; yet the people gave Perseus a royal welcome, and it was not long before they asked him to rule in Acrisius' place. At first Perseus was unwilling to do so, but at last he agreed.

It remains only for me to tell how Perseus gave the winged sandals back to Mercury, and the shield to Minerva – together with Medusa's head, for which he had no further use. Later he made peace with his great-uncle, King Prœtus, and when Prœtus died Perseus ruled over both parts of Argos. His reign was long and glorious, and he never forgot his promise to take Andromeda back to see her parents whenever he could be spared from his kingly duties. He had many more adventures in his long life, but never again did he have to face so great a danger as when he had set out in quest of the Gorgon's head.

You should have little difficulty in finding most of the characters of this legend in the star-maps on the next page and on page 48 and 49. The most conspicuous of them, oddly enough, is not Perseus, but Cassiopeia.

First, identify the Great Bear or Plough, which is an easy matter; most people know where to look for the seven famous Plough-stars. Then find the Pole Star, as shown in the map on the next page. Once this has been done, you can use Mizar – the second star in the Bear's tail – together with the Pole Star in order to recognise Cassiopeia. The proud queen does not look much like a lady in a chair, but her prominent W-shaped outline is very well marked.

Cassiopeia, like the Bear, never sets over Britain, and so may always be seen whenever the sky is dark and clear. Sometimes, as in January evenings, the constellation is almost overhead; even at its lowest, on summer evenings, it is always well above the

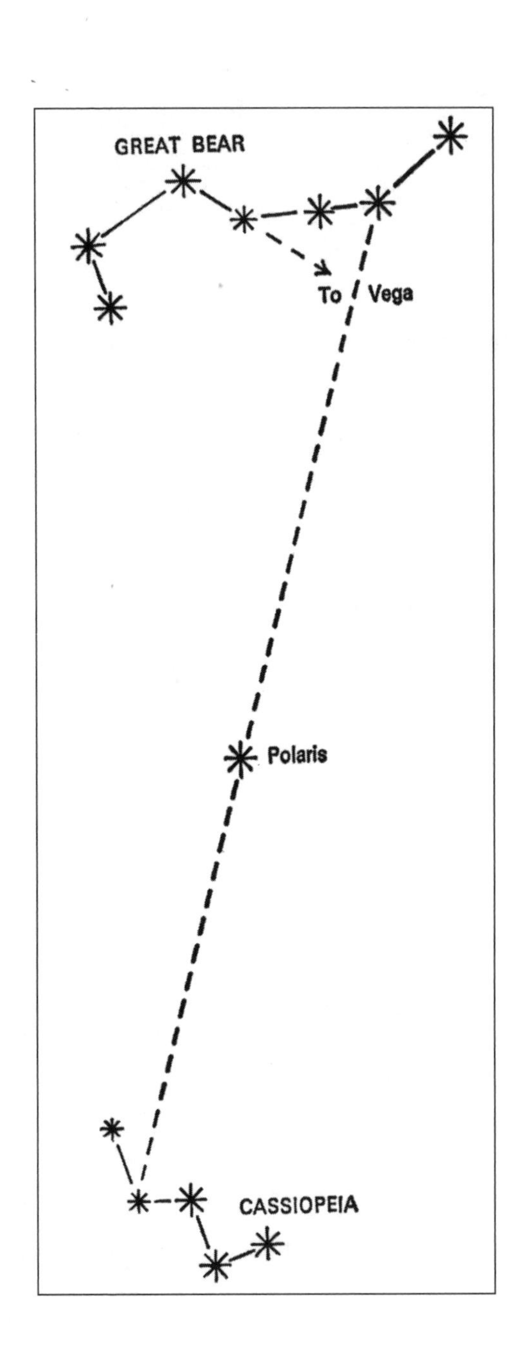
GREAT BEAR
To Vega
Polaris
CASSIOPEIA

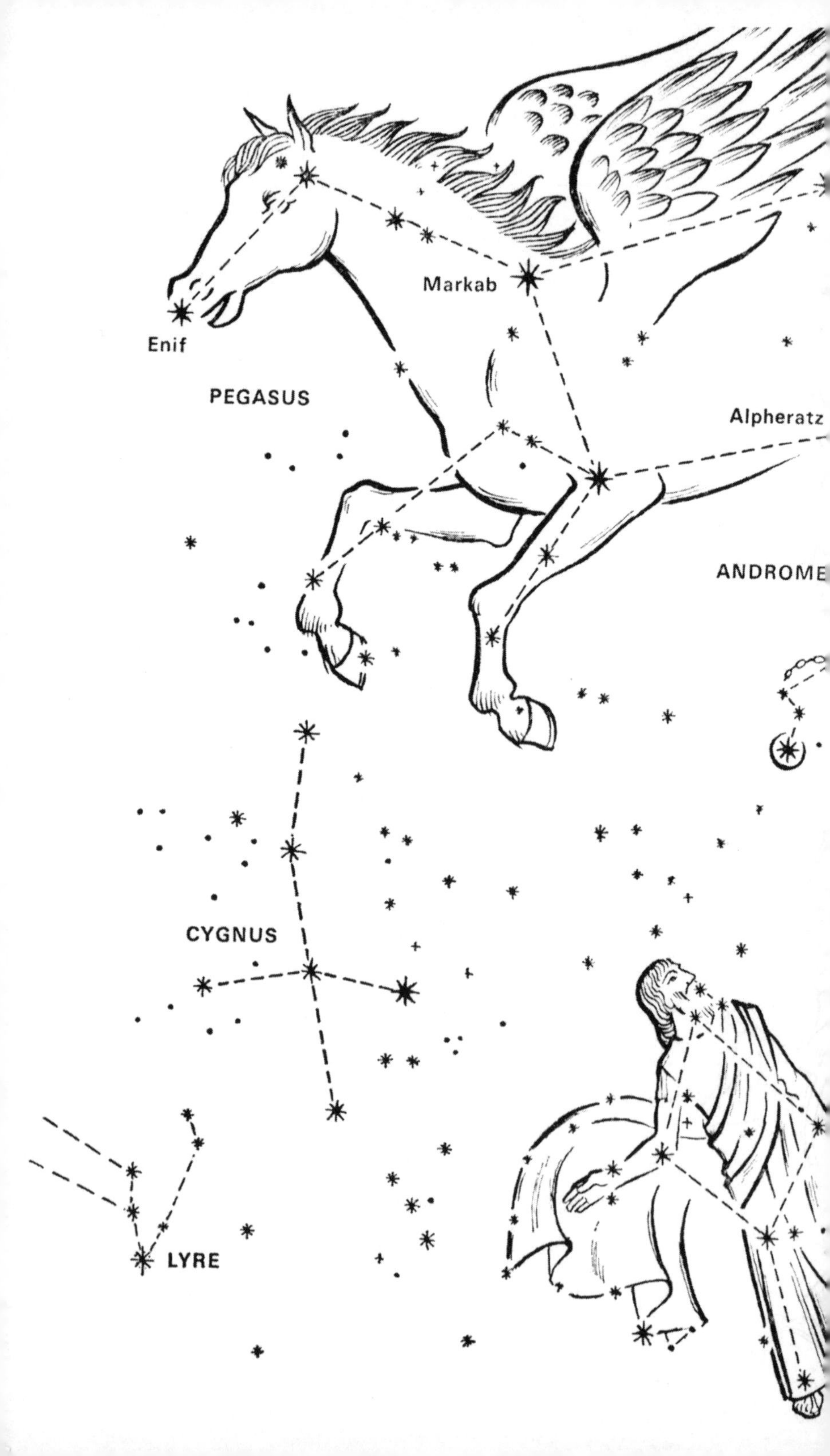

Markab
Enif
PEGASUS
Alpheratz
ANDROME
CYGNUS
LYRE

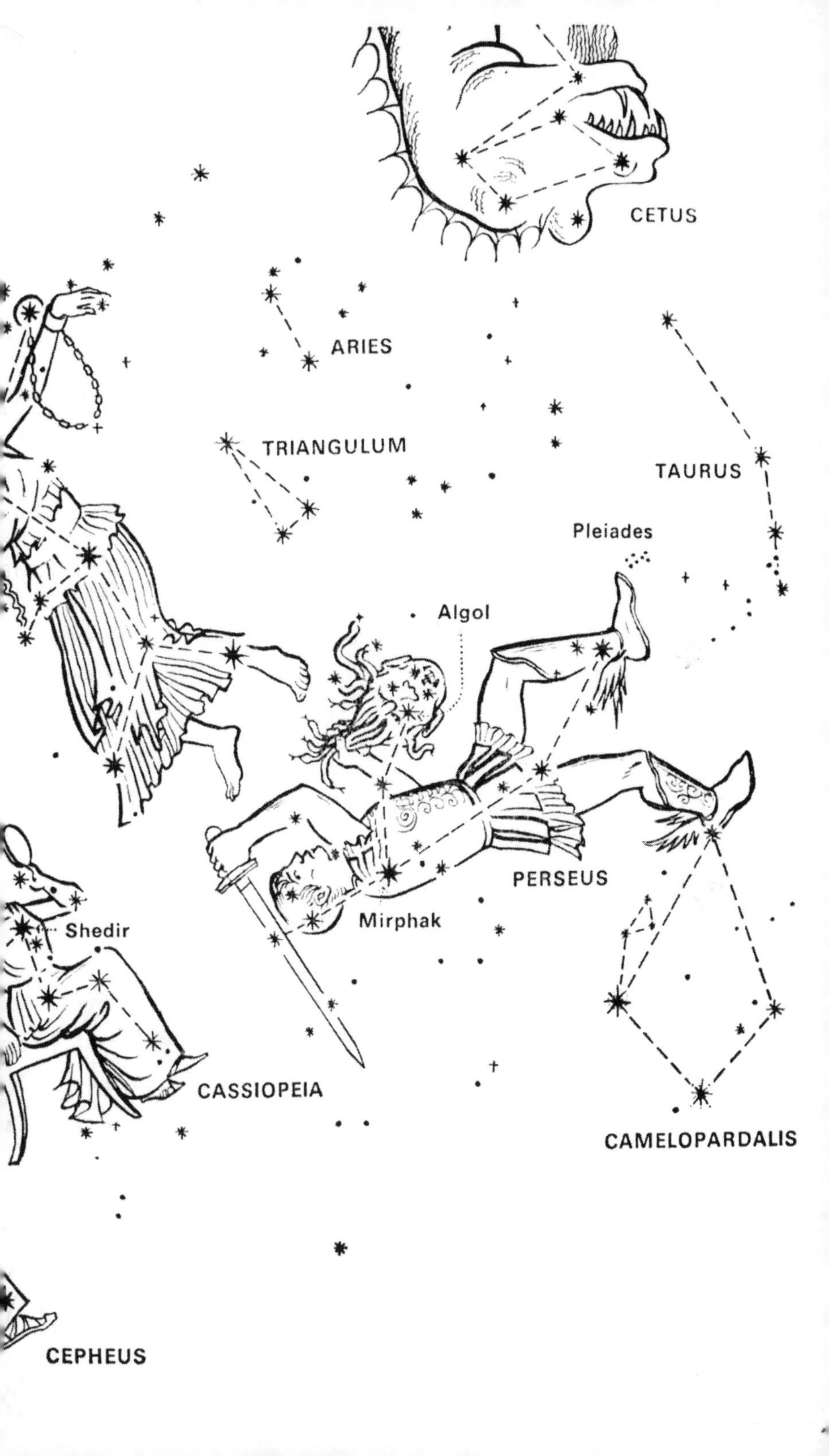
CETUS
ARIES
TRIANGULUM
TAURUS
Pleiades
Algol
PERSEUS
Mirphak
Shedir
CASSIOPEIA
CAMELOPARDALIS
CEPHEUS

northern horizon. Cepheus is not so bright, but can be found near Cassiopeia.

Now use two of the stars in the W of Cassiopeia to find the square of Pegasus. Pegasus, the flying horse, belongs to the next story, but one of the stars in the square belongs to Andromeda, and the rest of the constellation is made up of a line of fairly bright stars together with a good many fainter ones.

Take particular note of Alpheratz, which is about as brilliant as the Pole Star.

Beyond Andromeda lies Perseus himself, clearly recognisable, and with one prominent star, Mirphak. The Gorgon's Head is marked by a rather fainter star, Algol, which is of very unusual type. Every two and a half days it gives a long, slow 'wink', fading so much that it becomes quite dim when seen without a telescope. You will not be able to watch it fading before your eyes, but if you look at Perseus and find that Algol is not of its usual brightness you may be sure that a 'wink' is going on.

Modern astronomers have found out why this happens. Algol is made up not of one star, but two. One member of the pair is much brighter than the other, and the two are moving together round their centre of gravity, much as the two bells of a dumb-bell will do when spun round by their joining bar. It takes two and a half days for the pair to complete one turn – and when the fainter star passes in front of the brighter star, blocking it out, a 'wink' is the strange result.

It used to be thought that the ancient astronomers knew all about this, and so named Algol the 'Winking Demon' in remembrance of Medusa. Nowadays it is not believed that the old star-gazers realised how Algol behaves, but at least it is suitable for the Gorgon's Head to be marked by a winking star.

Lastly, what of the sea-monster? You can find him, too, marked in the star-maps as Cetus. He lies well below Andromeda, and it is surely right that he should be in the sky; he is not bright, but he is very large indeed. An imaginary line drawn from Mirphak through Algol, and continued for some distance, will lead you straight to him.

Such is the story of Perseus, one of the bravest and boldest of all the heroes in ancient legend. His courage never failed; not even the terrible monster sent by the sea-god could hope to overcome him.

3

The Flying Horse

You will remember King Prœtus, who was the great-uncle of Perseus and brother of the wicked King Acrisius. Prœtus was not really a bad sort of person; he ruled over that part of Argos which was called Tiryns, and his subjects were quite content to have him as their sovereign.

One day Prœtus had an unexpected visitor. A tall, handsome young man arrived at court, dressed in rich robes, and asked whether he could see the king. Prœtus gave orders for the stranger to be shown in, and looked at him in surprise.

'Who are you?' he asked. 'I can see that you are no penniless wanderer, and indeed you look like a prince.'

The young man bowed. 'Greetings, great king. You are right; I am a prince – Bellerophon, prince of Corinth. Yet I am a penniless wanderer too, in spite of my fine clothes, and I have come to ask whether you will allow me to stay here for a while.'

Prœtus looked quickly at his wife, Queen Anteia, who was sitting beside him. 'Hmm,' he said thoughtfully. 'It all seems very curious. If you are a prince of Corinth, why have you not come here with your servants and your horses? It seems very unlikely that a nobleman should arrive without money or friends. I am bound to say that I think you must have been guilty of some crime.'

'I have been guilty of some crime – but not by my own wish,' said Bellerophon quietly. 'I was taking part in some games when I threw a heavy discus, and accidentally killed one of my companions. I was about to marry a beautiful princess, but now she will have nothing more to do with me, and I have been driven out of my own country. If you will let me stay here, I will serve you faithfully.'

'I really don't know what to say,' muttered Prœtus. 'I can always make use of strong young men, and if your story is true I can see that you have been very unlucky, but how do I know that you are not telling me a pack of lies?'

Bellerophon flushed. 'Let me prove my honesty, sir. That is all I ask.'

'I believe he is telling the truth,' broke in Queen Anteia, and whispered in Prœtus' ear: 'Give him a chance to serve you, at least. He seems to be a loyal and brave person.'

Prœtus agreed, not without hesitation, and Bellerophon came to live at the king's court. The truth of the matter was that Queen Anteia had taken a great liking to him, and she could usually make Prœtus do as she wished. Unfortunately the queen was an evil and treacherous woman, and after a while Bellerophon offended her so greatly that she made up her mind to cause his downfall. So she went to Prœtus, and told him a great many wicked lies, which the king unhappily believed.

'I warned you about this young man from the very beginning,' he said irritably. 'It is really your own fault, my dear Anteia. I wish I knew what to do with him.'

'Kill him,' said Anteia, and clenched her hands. 'See that he dies. He deserves nothing better!'

But Prœtus, weak though he was, could not bring himself to order the death of a man whom he had treated as a guest. Anteia did all she could, but when she found that for once the king would not change his mind she persuaded him to send Bellerophon off on a mission to her father, King Iobates of Lycia, who, as we remember, had sent an army to help Prœtus in his war against King Acrisius. Bellerophon was sent away, taking a message to Iobates. He did not open the sealed message, which was a pity, since Prœtus had told King Iobates that the man who delivered the letter was to be killed as soon as possible.

Prœtus had certainly behaved in a very shameful manner; the only excuse for him is that he really believed Bellerophon to have been guilty of all sorts of crimes, and he could hardly be expected to know that his wife had tricked him. Meanwhile, Bellerophon arrived at the court of King Iobates and handed over the letter, quite unaware that there was anything the matter.

Iobates read what Prœtus had written, and looked stern. 'Remarkable,' he thought to himself. 'This young fellow does not look at all like a scoundrel – but my daughter Anteia says he is the blackest of villains, so I suppose there is no doubt about it.' Aloud he said: 'Greetings, Bellerophon. You know what is in this message, of course?'

Bellerophon shook his head. 'I have not opened it, sir. It was not meant for me to read.'

'I can well imagine that,' said Iobates meaningly. A plan had come into his head; he did not want to kill Bellerophon outright, but there was a much better way. 'I hear that you are famous for your courage and your skill. I have been having a great deal of trouble lately, and King Prœtus seems to think that you are the only man who can help me.'

Bellerophon drew himself up. 'I am honoured, your majesty. Tell me what is to be done, and I will do my best.'

Iobates paused. 'It is a dangerous task,' he said. 'No, no – I really cannot ask it of you. Forget what I have said.'

'I am not afraid,' said Bellerophon. 'Tell me, and I am at your service.'

'You are sure? The risk will be great,' said Iobates, and shook his head sadly. 'I will tell you my problem, but if you do not feel that you can help me – well, I shall understand. My enemy, the king of Caria, has made a pet of a terrible monster known as the Chimæra. It has the head of a lion, the body of a goat, and the tail of a huge snake; it breathes fire, so that nobody can possibly go near it. I am very much afraid that as soon as he has gathered an army, the king of Caria will march against me, sending his monster on ahead to burn up my towns and scorch my people.'

King Iobates, of course, was doing his best to trick Bellerophon into going off to kill the fire-breathing dragon; it seemed much more likely that the dragon would kill Bellerophon instead, which was what both King Iobates and King Prœtus wanted. But Bellerophon did not need persuading. 'I will try to kill this Chimæra for you,' he said, and drew his sword, feeling the blade thoughtfully. 'If I cancreep close enough without being being burned to a cinder, I am sure that I will be able to give the monster a fatal wound, but it will be very hard to avoid being scorched up. Does the Chimæra breathe fire all the time?'

'All the time, even when it is asleep,' said King Iobates grimly. 'I wish I could give you some advice, but there is really very little that I can tell you. The only man wise enough to help, so far as I know, is Polyeidus.'

'Polyeidus? Who might he be?'

'A strange old fellow, who lives alone some way out of the city,' explained King Iobates. By this time he was rather regretting that he had persuaded so brave a youth to go off on a hopeless errand, yet he could not quite bring himself to believe that his daughter Anteia might not be telling the truth. 'You will find Polyeidus easily enough; take the road that leads out to the west, and after you have walked for an hour or so you will come to the log hut in which he lives. If he cannot advise you, you had much better come straight back here and forget all about the Chimæra.'

Bellerophon had no thought of doing any such thing, and he made his way to Polyeidus' cabin as quickly as he could. It was evening by the time he arrived, and he found the wise man sitting comfortably outside his hut, looking at the setting sun. Polyeidus was very old, but his eyes were bright and clear, and he did not seem at all surprised when Bellerophon came up and told his story.

'I've heard of you, young man,' said Polyeidus. 'I need not tell you that you are in serious danger – perhaps, indeed, in more danger than you realise. There is only one way in which you will be able to destroy this fire-breathing monster. If you attack it from the ground you will certainly end up by being scorched to a cinder, which will do nobody any good.'

'I don't want to be scorched to a cinder,' said Bellerophon firmly. 'In fact, there is nothing that would appeal to me less. What's the best thing to do, then? I can't fly, so I suppose I must simply take my chance. The Chimæra must go to sleep sometimes, and I may be able to steal up before it wakes.'

'That would be impossible. No, there is only one thing for you to do.' Polyeidus stood up, and pointed towards the sky. 'You must catch the flying horse Pegasus, and ride upon his back.'

Bellerophon gave a gasp. Of course he had heard of Pegasus, the wonderful white horse with the flowing wings, who lived in the air and sometimes came down to drink the pure water from the mountain pools. But he knew that Pegasus was immortal, and that no human had ever succeeded in touching him, let alone riding on his back. 'Catch Pegasus!' he repeated, looking at the wise man in alarm. 'I shall never be able to do that. Why, I should need the help of the gods themselves.'

'That might be arranged,' said Polyeidus, and took Bellerophon by the arm. 'Minerva, the goddess of wisdom, is always ready to give her aid to brave men such as you. I have some magic powers, as you know, so I will see what can be done. Lie down here and go to sleep. When you wake tomorrow morning, I may have more to tell you.'

Bellerophon felt very tired, so he did as Polyeidus told him, and stretched himself out on the soft grass. No sooner had he closed his eyes than he was sound asleep, and he did not wake until the sun had risen next day. As he sat up, he saw Polyeidus by his side, holding out a golden bridle. 'You are fortunate,' said the old man, and gave a soft chuckle. 'Minerva came in the hours of darkness, and she brought this bridle with her. All you have to do now is to find Pegasus, creep up behind him, and slip the bridle over his neck.'

Bellerophon felt the bridle, which was very beautiful indeed, and glittered brightly in the morning sunlight. 'I owe you my deepest thanks, Polyeidus, but I have no idea where Pegasus may be found. He may be hundreds of miles away, and I cannot follow him through the air!'

Polyeidus clicked his tongue. 'Don't be so gloomy,' he said, in a rather cross tone. 'Pegasus may be a remarkable horse, but he has to drink sometimes, and he lands beside one of the mountain springs. I believe that if you go to the spring on the Acropolis of Corinth, which you know well, you will not have very long to wait. Take this lump of lead, by the way; you will need it.' He held out a large piece of metal, and Bellerophon, feeling very puzzled, took it. 'I have done all I can, and from now on you must depend upon your own courage and common sense.'

This was true enough, so Bellerophon thanked the wise man once more, and said goodbye. Then he made his way to the mountain spring, which was a great many miles off. It took him more than a day and a night to walk there, but at last he arrived, and sat down to wait. He was quite alone; the spring bubbled, and the water glistened in the sun's rays, while all around him were the snow-capped mountain peaks and the rich green woods below.

Bellerophon stayed there for another whole day and night, sleeping as little as possible, and wondering how long he would have to wait. He closed his eyes from time to time, but did his best to keep wide awake, listening for any sound which would tell him that Pegasus was near.

Suddenly, just after dawn on the second day, he heard a soft rushing noise in the distance. Quickly Bellerophon hid himself behind a rock, and gripped the golden bridle. Could this be Pegasus? It sounded like some great flying creature, and in a moment more he saw that a graceful animal had dropped from the sky and was trotting towards the mountain pool. It was Pegasus, proud and beautiful, wilder than the hills, and yet gentle too. His wings dipped as he bent his head, drinking deeply from the clear waters.

This was Bellerophon's chance. Swiftly he darted forward, thrust the bridle over Pegasus' neck, and leaped upon the horse's back, clinging on ready for anything which might happen. You can imagine that Pegasus was startled, and gave a loud whinny, rearing up so as to throw off this impolite stranger who had jumped upon him.

'I mean no harm,' called Bellerophon, wondering whether the magic horse could understand. 'I am your friend, and I have been sent here by the Goddess Minerva ...'

There was not time for more. With a swoop and a rush Pegasus was off, soaring through the sky, twisting and turning as he did his very best to shake free. It was a wonder that Bellerophon did not fall off, and the mad flight made him feel sick and giddy; now he could see the earth, now the sky, now the earth again. Somehow he kept his place, and after a while Pegasus became quieter. Bellerophon patted him gently, and the magic horse turned his head.

'I mean you no harm,' said Bellerophon again, and this time Pegasus gave a soft whinny instead of a scream. 'Let me fly upon your back, and help me to keep my promise to King Iobates.'

Pegasus seemed to know what Bellerophon meant, and for the next hour the two flew here and there, diving and swooping until Bellerophon was sure that the flying horse would do as he was asked. It was a truly wonderful feeling; no travel on the ground could be nearly so exciting as this ride through the sky, and already it seemed that Pegasus thought of Bellerophon as a master as well as a friend.

At last Bellerophon felt that he was ready to start his real task, so he turned Pegasus until his head pointed towards the kingdom of Caria. 'This is the way we must go,' he said, and settled himself down. 'We must fly to the land of King Iobates' enemy, where we will find the fire-breathing Chimæra. Make haste!'

Pegasus whinnied again, and began to soar along as quickly as the wind. Before long they had left the land of Corinth, and were passing over Caria, which Bellerophon had never seen before. There were towns and villages here and there, set among the mountains, but at last they came to a broad plain, in the middle of which lay the enemy king's chief city. It was here, some way from the royal palace, that they first caught sight of the Chimæra.

The first thing that Bellerophon saw was a puff of smoke, together with a red glare which looked like a large fire. He brought Pegasus down until they were no more than a few hundred feet above the ground, and then he could make out what appeared to be a huge grey cloud. In a moment or two the smoke parted, and there, underneath, was the Chimæra. It was a fearsome monster indeed; as Iobates had said, it had the head of a lion, the body of a goat and the tail of a snake. It was huge, too, and its roars could be heard even above the swish of Pegasus' wings. Each time it breathed there was a red glow, and the smoke swirled out across the plain. All around, the grass was blackened and burned; nothing could live near the Chimæra.

'It must have been made by King Pluto himself,' thought Bellerophon, and felt thankful that he was safe in the air. 'Keep high, Pegasus, or we shall be scorched. Stay still, now!'

He brought Pegasus to a halt until they were hovering high above the Chimæra, drawing out his bow and arrow and taking careful aim at the monster's head. Then he let the arrow loose, and waited. His aim was true; the arrow struck the Chimæra, but for all the harm it did – well, it might have been a pebble. The arrow simply bounced off and fell to the ground.

Again and again Bellerophon aimed, but with no better luck. Once it even seemed that he hit the Chimæra in the eye, but the monster took no notice at all, and Bellerophon began to wonder whether it, like Pegasus, could be immortal. 'What shall I do?' he said aloud. 'My arrows are useless, and I shall certainly not be able to thrust my spear into the creature even if I can get close enough to try. Polyeidus never warned me about this!'

Then he remembered that the wise man had given him a lump of lead, saying that he would need it. There was no sense in dropping the lump on top of the monster, so what could Polyeidus have meant? Bellerophon thought hard, and then he saw what he must do. He drew out the lead, and managed to fix it upon the tip of his spear, making sure that it would not drop off.

'We must swoop down,' he said, and turned Pegasus' head towards the ground. 'Wait until I give you the signal, and then fly past the Chimæra as closely as you can. Be quick, or we shall be caught in its fiery breath!'

Pegasus seemed to understand, and Bellerophon brought him down until they were almost within range of the smoke and flames sent out by the Chimæra. Bellerophon paused, and waited for his chance. The monster could breathe out fire only every half-minute or so, and this might give Pegasus just enough time to fly past.

The flames broke forth, and the smoke from the monster's mouth puffed over the plain. Then Bellerophon gave a shout, and sent Pegasus onward until he was within a few feet of the terrible jaws. The Chimæra reared up, and gave a snarl like thunder, drawing back so as to scorch up these strange flying creatures, but Bellerophon was too quick. He leaned forward, almost overbalancing in his haste, and thrust the lump of lead straight into the Chimæra's mouth.

The huge jaws snapped shut, and Bellerophon tugged at the bridle, sending Pegasus shooting up into the sky as fast as his wings would take him.

From below there came another roar – not a snarl, this time, but a hideous choking sound. The Chimæra breathed out, and the flames melted the lump of lead fixed in its jaw. As Bellerophon turned and started down, he could see that the monster was dying; the melted lead had blocked the Chimæra's mouth, so that no air could pass through.

Pegasus flew around, circling above the Chimæra, while Bellerophon watched. There was yet another roar, but now weaker, and when the Chimæra tried to breathe out there was no red glare or cloud of smoke. All that came was a small puff, and the monster fell, its lion's head dangling and its snake's tail thrashing about. Slowly the movements became less and less, and at last the Chimæra moved no more. It was quite dead, and could do nobody any further harm.

'Polyeidus was right,' thought Bellerophon, and patted Pegasus affectionately on the neck. 'All the same, I could never have killed the Chimæra without your help, my wonderful horse. Let us go down and make sure that the monster has no life left in it.'

He pointed to the ground, but for some reason or other Pegasus did not seem willing to go. Bellerophon looked back at the city, and then he saw the reason why Pegasus had no wish to land. A whole army of warriors was streaming out on to the plain, and most of the men were looking upward, shaking their fists and shouting loudly. Next moment an arrow flew skyward, and Bellerophon lost

no time in putting Pegasus in flight once more, soaring up well out of range.

Bellerophon would have liked to have taken a closer look, but it was clear that the king of Caria was furious at the death of his Chimæra, unpleasant pet though it must have been. Without it, he could hardly hope to make a successful attack on King Iobates, and Bellerophon fancied that he could see him at the head of his troops, waving his fists more angrily than any of the soldiers. There was no sense in waiting, so Bellerophon turned Pegasus to the east

Bellerophon attacks the Chimæra

and started on the long flight back. After many hours had passed, he was once more over Lycia, King Iobates' country, and within sight of the royal palace.

You can well picture that Iobates and his courtiers were surprised to see Bellerophon alive and unharmed, and even more surprised to find that he was riding on the enchanted horse. No sooner had Pegasus landed than Bellerophon dismounted, walked up to the king, and gave a low bow. 'I have carried out your orders, sir. The Chimæra is dead.'

'Dead! The Chimæra?' stammered Iobates. 'But – but that is impossible. How could you manage to kill such a creature?'

'With the help of the wise man Polyeides, the goddess Minerva, who gave me a golden bridle, and my faithful four-footed friend here,' said Bellerophon modestly. 'You will certainly have no more troubles from the king of Caria. I'm sure that he believes the gods themselves to be fighting against him.'

'He may well be right,' muttered Iobates, and looked hard at Bellerophon. Surely this gallant young hero could not be the scoundrel whom his daughter Anteia had described? Yet still he could not believe that Anteia was lying, and he decided upon one more test. 'My good Bellerophon, you have done well. The thanks of myself and my people are due to you, and we will never forget your bravery. While you still ride Pegasus, there is one more task that I would ask you to do – quite a simple one, I assure you; not nearly so dangerous as tackling that dreadful Chimæra.'

By now Bellerophon felt that he was more than a match for any man, or for that matter any creature that had ever walked on earth. 'I am at your service, sir. Tell me what it is that you want done.'

Iobates thought quickly. 'As you know,' he said, 'I am king over all Lycia, and my subjects are loyal and obedient. I think I can say that I am a kindly ruler, and I have always been ready to listen to any complaints, so that I may put them right if I can.'

Bellerophon bowed again. 'So I have always heard, sir. For myself, I am grateful to you for having made we welcome here.'

'Unfortunately,' went on Iobates, 'there are two wild tribes in the northern part of the country who are always giving trouble, and to make matters worse there are some pirates sailing up and down the coast, sent there by the king of Caria to attack my ships. Up to now we have never found any way of dealing with them, but if you can kill a Chimæra it should not take you long to finish off some disobedient tribesmen and a gang of ruffians on the sea.'

You may be sure that Bellerophon made no objections, and after he had eaten and slept, not forgetting to give Pegasus plenty of food and a comfortable stable for the night, he set off again. All the townspeople gathered together to watch him go, and as Pegasus mounted into the air the people let out a cheer. Even King Iobates waved his hand, and once more he felt disturbed. As soon as Pegasus was out of sight, the king called his palace soldiers together.

'Listen to me carefully,' he said, as the soldiers drew up in line and saluted. 'I must tell you that my daughter, Queen Anteia – the wife of our friend and ally King Prœtus, ruler of Tiryns – has sent me a letter in which she tells me that Bellerophon, whom you have just cheered, is nothing better than a rascal who deserves death. I admit that I find this very hard to believe, particularly as Bellerophon has done us all such a great service by killing the fire-breathing monster which the king of Caria meant to send against us, but I cannot doubt my daughter's word.'

There were murmurs from the soldiers, and the chief officer stepped forward. 'Surely there must be some mistake, my lord king.'

'That is what I would like to believe,' said Iobates, 'but – well, I confess that I don't know what to think. Be ready, all of you. I have no doubt that Bellerophon will soon be back, having done all that I have asked. As soon as he lands, seize him and bring him before me. I must know the truth!'

With these words, King Iobates went back into his palace, still thinking deeply. He had half a mind to send a letter to Prœtus, but no messenger could travel nearly so fast as Pegasus and there would be no time to receive an answer.

The day passed slowly. So did the night, and then the following day. There was still no sign of Bellerophon, and King Iobates began to wonder if, after all, he had met with some accident. He sat in his throne-room, and waited with as much patience as he could manage.

Suddenly he heard a noise from outside, and he walked over to the window, peering out to see what was happening. Before he could do more than glimpse the crowds outside, the door burst open and Bellerophon strode in, his eyes flashing and his sword held in his right hand. 'What is the meaning of this, my lord king? I have carried out your commands; I flew over the tribesmen and dropped huge stones upon them until they surrendered, and then I sank the pirate ships so that they can trouble you no more. Yet when I return, your men set upon me with their swords, and it is a wonder that I did not have to kill several of them.'

Iobates shrank back; he did not like the look of the sword in Bellerophon's hand, and he fumbled for the letter which had been the cause of all the trouble. 'Read this,' he said. 'Read, and tell me whether what my daughter says is true or false.'

Bellerophon snatched at the letter, and read it through. 'These are black lies, my lord – so black that I will not even take the trouble to deny them. I see now why you sent me to track down the Chimæra, and why your men attacked me when I came back. You are ungrateful, King Iobates. I will leave your court, and you will see me no more.'

Iobates stepped forward, and threw out his arms. 'I believe you, young man. No scoundrel could act as you have done, and I must realise that my daughter Anteia is a very wicked woman, much though I dislike the idea.' He laid a hand on Bellerophon's shoulder. 'Stay here, I beg you. You shall have a place of honour at my court, and I know that my younger daughter Philonoe, who is even more beautiful than Anteia, will be anxious to marry you. I have behaved very badly, but I am truly sorry, and I ask forgiveness.'

By this time Bellerophon's anger was beginning to pass away, and he saw that King Iobates really had been in a very difficult

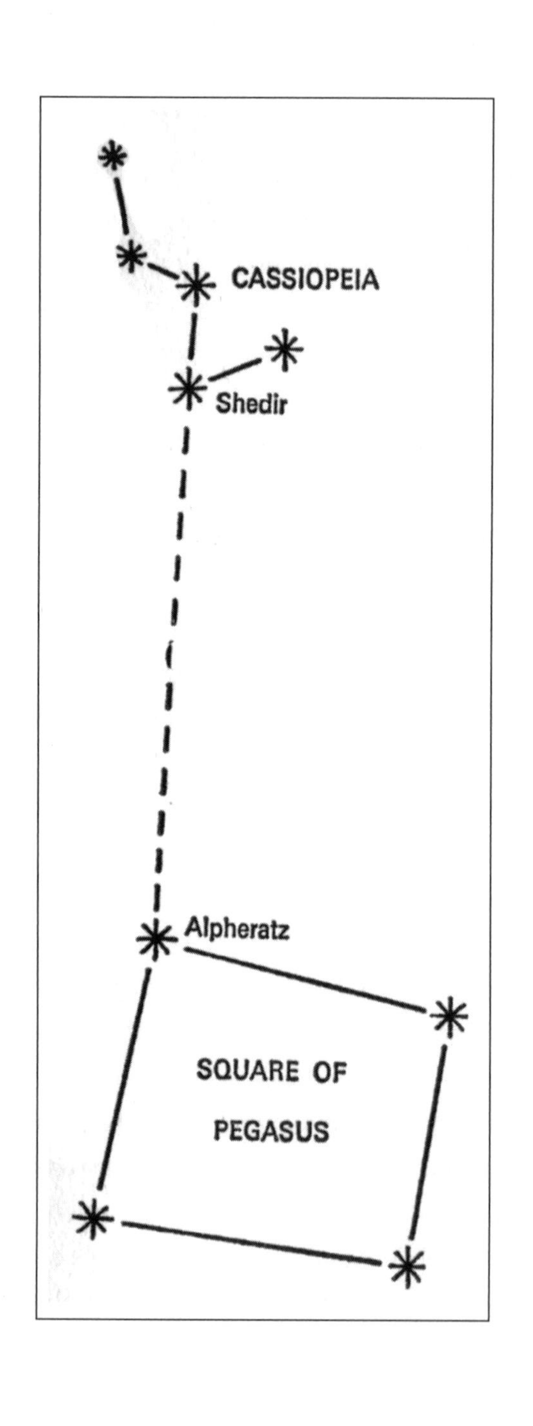
CASSIOPEIA
Shedir
Alpheratz
SQUARE OF
PEGASUS

position. So he agreed to stay at the royal court; and when the Princess Philonoe came to him, he fell in love with her at once. The wedding took place a month later, and all the people of Lycia came to it, cheering loudly and waving to the man who had freed them from the threat of the terrible fire-breathing Chimæra.

You will not find Bellerophon in the sky, but Pegasus is there, and you can see him on any clear evening throughout the late summer and the autumn.

The best way to find him is to use the W of Cassiopeia as a guide. The illustration opposite will show you how to find Cassiopeia, by using Mizar in the Great Bear, and the Pole Star in the Little Bear, as pointers. Now trace an imaginary line from two of the W-stars, as shown here, and you will come to a conspicuous 'square' of four fairly bright stars. As we have seen, one of them, Alpheratz, is in Andromeda; the other three are in Pegasus, and the pattern is always known as the Square of Pegasus. The other stars of the Flying Horse lie around about, and you should have no difficulty in finding them.

It is quite true that Pegasus is in no way as striking as might be thought; most people expect the Square to be smaller and brighter than it really is. Neither will you see it all the year round, since it sometimes drops below the horizon. All the same, Pegasus is clear and unmistakable, and once you have found him you will not forget him. Every autumn evening he shines down, and with a little imagination you will be able to trace the starry outlines of the swift-moving, graceful Flying Horse.

4

The Story of Arion

The Greeks, as you know, were very fond of music. They were splendid singers, and they were also skilful players of the stringed instrument known as the lyre. Not many lyres are to be seen nowadays, but there were many of them in Ancient Greece.

Of all the musicians, the best, as everyone agreed, was Arion. This was not surprising, since it was said that Arion was related to the sea-god Neptune; if he had been an ordinary man, he could not have produced such wonderful songs. Nobody seemed to be quite sure, but in any case Arion was famous all over the Mediterranean world.

Arion had been born on Lesbos, one of the Greek Islands, but when still a boy he had gone to the court of King Periander of Corinth (we remember that Bellerophon, too, was a Corinthian, though as a matter of fact Arion lived long after the time when Bellerophon had mounted the flying horse and killed the

Chimæra). King Periander had been very glad to welcome Arion, and had made him his friend as well as his court musician. Every evening the people flocked in to hear Arion's songs and tunes, and the kingdom would have seemed very dull without him.

One day Arion heard that there was to be a great musical festival at Taenarus. Taenarus, which we now call Taranto, was not in Corinth, nor even in Greece; it lay on the island of Sicily, at the end of the boot-shaped country of Italy. This meant that if Arion were to take part in the festival, he would have to undertake a fairly long sea-voyage, and King Periander was not at all anxious for him to take the risk.

'It's a long way,' he said doubtfully, pulling his beard and looking hard at Arion. 'Oh, I know that our ships go across to Sicily every month or so, but – well, there are all sorts of dangers. There are pirates, for instance. We've cleared as many of them as possible, but there are still a great many pirates' ships in the Mediterranean.'

'I'm not afraid,' said Arion. 'Besides, I am not a coward, even though I am a musician and not a fighting-man. You are my friend as well as my king; I beg you to let me go. It is not as though I shall be away for long.'

'Oh, very well,' grumbled Periander, pulling his beard harder than ever. 'I admit that I don't like the idea at all, but as you are so anxious to take part in this musical festival I can see it would be quite wrong of me to stop you. Come back as quickly as you can, Arion. My court will seem very silent and gloomy while you are gone.'

So Arion made his departure, and boarded one of the king's ships which was due to sail for Taenarus. It had been many years since he had left Corinth, and although he liked his home, and was very fond of King Periander, it was a pleasant change to be on his own for once. Of course he was very popular with the sailors, and all through the voyage he entertained them with his music, so that it seemed a very short time before they saw the coast of Sicily ahead of them.

The town of Taenarus was brightly decorated with flags and banners, since all the best musicians in the world were coming there to take part in the festival. There were concerts, and musical turns of all kinds, in which Arion took a leading part; there were even moments when all the other performers stopped, and sat silently while they listened to Arion's splendid singing and playing. Then came the competitions, in which all the players entered. There were prizes for the best songs and tunes, but it soon became clear that nobody could hope to win against Arion.

'I really think that I should take no more prizes,' said Arion at last, looking rather ashamed. 'It is very kind of you all to do me so much honour, but I feel it is wrong for any one man to win too much. If you wish, I will share out the prizes I have already won.'

'No,' said the chief judge, shaking his head. 'This is a fair competition, Arion, and you are winning all the prizes simply because you sing and play better than anybody else.' The other competitors standing around nodded in agreement. 'Besides, you are giving us great pleasure with your music, so that we are having our prizes too – even if they are not in the form of money and jewels. Let us continue.'

So they went on and on, and Arion won more and more prizes. Again he offered to share them out, but the people refused to let him do so, for, like the judges, they were delighted just to listen to Arion. So when the festival was over, and Arion was ready to keep his promise and sail back to Corinth, he had a very valuable load to take with him.

Up to now, everything had been very pleasant. Arion had enjoyed the festival; he had made friends with the people of Sicily, and he expected to find the homeward voyage as calm as his outward journey. Unfortunately, the ship which had brought him to Sicily had already sailed back, and Arion had to go on board another vessel. This did not worry him in the least, but he would have been very anxious if he had known what the captain of the ship had in mind. The captain, I regret to say, was a villain. He knew that Arion was bringing a large sum of money, together with rich jewels and other prizes, and he meant to get them for himself.

For the first day or so all seemed to be well, and Arion sang and played as before, while the sailors listened and the ship glided on towards Corinth over the calm waters. Then, when they were out of sight of Sicily and were still a long way from Greece, the captain signalled to his men. Arion, sitting by the side of the ship, was fingering his lyre and looking across at some dolphins who were playing in the water not far off; as he touched his lyre, the dolphins paused to listen, and we may suppose that they even beat time in the water with their tails (dolphins, of course, are not fishes; they are animals, and very intelligent ones at that; in fact, a dolphin is just as clever as a dog).

Suddenly, Arion gave a cry of alarm. He felt himself seized, and pulled roughly away from the side of the ship, his lyre falling to the ground. He struggled, and found that he was being tightly held by two sailors, while the captain stood by and looked on.

'What is the meaning of this?' said Arion angrily, and went on struggling. 'Captain, order your sailors to let me go. I have done nothing wrong!'

'That is true,' said the captain slowly, 'and I admit that I feel rather ashamed of myself, but I am a poor man, and I really cannot let so splendid a chance pass by. You have won enough money to make me and my crew rich for the rest of our lives; once we have taken possession of your winnings, we can settle down on shore and live in comfort. I am sorry to have to kill you, but there is nothing else that I can do.'

Arion felt frightened. Remember, he was not a hero such as Perseus and Bellerophon had been; though he did not lack courage, he was a man of peace, and he had never had to take part in a fight. Besides, he was alone. There were more than a dozen sailors on board the ship, and it would have been hopeless for him to try to do battle against them.

'That would be a wicked thing to do,' he said, and looked the captain straight in the eye. 'As you know, I said that I would share my prize-money with the other musicians at Taenarus. Very well; I can see that you are greedy, and that I am in your power. If you will spare my life, I will let you have all the money.'

The captain looked uncertain, but then one of the sailors – a broad, black-bearded fellow with a villainous expression – whisphered in his ear. 'Don't listen, captain. Arion will never keep silent, and when he gets back to Corinth, his friend the king will have us all arrested. If we're going to rob him, we must make sure that he is never able to tell the king what has happened.'

'You are right,' said the captain, though he still felt rather uncomfortable about it all. 'We wish you no harm, Arion, but you must realise that we can take no chances; after all, you are an important person in Corinth, while we are nothing more than common sailors. I fear that you must die.'

Arion looked out across the sea; the water was a lovely blue, and the dolphins were still playing, their silver bodies gleaming in the sunlight. He felt very sad that his life was coming to an end, but he knew that there was no way in which he could make the captain change his mind. 'Very well,' he said, after a long silence, 'I suppose I cannot complain; I have had a very happy life, and at least you will kill me quickly. Let me sing one last song, and then I will be ready.'

'Well, there can be no harm in that,' said the captain, and signalled to the sailors to let Arion go. 'Sing, then. When you have finished, jump into the sea. You will drown, of course, but drowning is not a painful death, and it is the best I can offer you. Take your lyre.'

Arion picked up his lyre, and felt the strings. What should he play? This would be the last time, so it would be only right for him to sing a hymn in honour of Apollo, who was the god of music and the arts. He held the lyre and began.

Of all Arion's songs, this was surely the most wonderful. The sailors listened, almost forgetting that they were about to kill this great musician; the dolphins clustered round the ship, and even the seagulls swooped down, perching on the masts. The waves themselves seemed to become silent, and nothing could be heard except Arion's voice and lyre. The captain was almost ready to burst into tears, but now that he had gone so far he knew that his crew would never let him draw back. Still the song went on, and the ship glided gently through the waters without making a sound.

The last note died away, and the silence was complete. Then Arion turned, and held out his lyre. 'You have heard my song, captain. For the last time, will you spare my life?'

The captain looked at his men, and then, slowly, shook his head. 'I would like to spare you, but I cannot. Jump into the sea, and make an end of yourself.'

Arion stepped towards the edge of the ship, and gazed at the deep blue water. Then he jumped, and found himself swimming, his lyre still in his hand. The captain did not wait; as soon as Arion was gone, he ordered the crew to make ready, and the wicked sailors made off as quickly as they possibly could, leaving Arion to die.

For some minutes Arion swam along; but he knew that he could not hope to reach the shore; he was not used to swimming, and in any case the coast was many miles away. He lost hold of his lyre, which floated off. Then, suddenly, he saw that one of the dolphins had left its companions and was coming towards him. As the friendly sea-animal approached, a new hope came into Arion's heart. 'Will you let me ride upon your back?' he asked, wondering whether the dolphin would understand.

As I have told you, a dolphin is a very clever creature indeed, and this particular dolphin must have been cleverer than all the rest. It swam up, and waited while Arion settled himself on its back. Then it turned its head, as though asking where it should swim. 'I am bound for Corinth,' said Arion, 'but we must not go too near the ship, or else the sailors will see me and make new efforts to kill me.'

Again the dolphin seemed to understand, and began swimming off to the east, cutting through the water more quickly than any ship could sail, and making sure that Arion's head and shoulders were kept quite dry. As they passed the lyre, which was still floating, Arion managed to pick it up, and once more he started to play, though he dared not sing too loudly in case the wicked sailors should hear and try again to kill him.

It was certainly a very strange way of making a voyage, but it was also a very pleasant one, and before long they passed the ship, keeping well away from it. No vessel could travel nearly so fast as the dolphin, and Arion found that he was not at all tired; once they were many miles away, he began to play again, and then the dolphin seemed to swim even more quickly then before. So they went on and on, until at last they came within sight of a shore that Arion knew to be Greece.

The dolphin slowed down, and swam gently into the shallow water. Arion slipped off its back, and patted its head. 'I am safe now,' he said, 'and I shall be at the king's court long before the sailors can arrive. You have saved my life, and I am truly grateful. Will you come

Arion riding the dolphin

with me to the royal palace? If so, I can promise you that you will be given all that you could wish for, and that you will be taken care of for ever.'

There are some people who say that the dolphin did indeed go with Arion, but died after reaching Periander's court. This is not true. A dolphin belongs to the sea; it loves the waves, the freedom and the open waters. So the dolphin indicated that it would not come, and as soon as Arion had reached dry land the friendly creature turned and swam away, looking back once or twice to make sure that his passenger really was safe and sound. Arion stood on the shore and waved till the dolphin had been lost to sight; then he turned, and started to make his way to the city of Corinth.

Now that he was back in his own country, he was out of all danger, and he had no difficulty at all in reaching Corinth, but you can well imagine how surprised King Periander was when Arion appeared, wet and dirty, and without any of his rich prizes. 'I told you that this sort of thing might happen,' said the king crossly. 'You should never have gone.'

'I met no pirates,' said Arion simply. 'The trouble was due to our own sailors, I fear. Before long they will be back, and I am sure they will tell you all manner of lies.'

'I see,' said the king grimly, after he had heard the whole story. 'As soon as the captain and his men arrive, I will order them to be brought before me. You can hide behind a curtain, and listen to what they are saying.'

Sure enough, the ship came into port on the following day, and the captain was brought to the palace, together with his crew. King Periander pretended to be very worried, and asked what had happened. 'You should have had Arion with you,' he said 'Where is he? Do not tell me that he has decided to stay in Sicily after all, in spite of his promise that he would return as soon the festival was over?'

'That is not so,' said the captain. He had his story all ready, and was sure that the king would believe him. 'Arion came on board, leaving his prizes behind so that he could collect them later. Most unfortunately we ran into a terrible storm, and Arion fell overboard before we could catch him. Of course we stopped, and managed to pick him up, but by that time he was dead. There was nothing we could do except give him a proper burial at sea,' he added, feeling that the whole story sounded very convincing.

'You are sure? And you really buried him at sea?' asked the king.

'Naturally, my lord. We put his body into a wooden coffin, and lowered it into the waves,' said the captain, feeling even happier. 'I can think of no better resting-place for a singer so wonderful as Arion.'

Periander made a signal, and the curtains parted; Arion stepped out, and the captain gave a shout of alarm, shrinking back. 'Arion! No, it can't be true. I must be dreaming!'

'I am no dream,' said Arion. 'You thought that you had left me to drown, but I was rescued by a dolphin, who took me on his back and brought me safely to Greece. What have you to say to that, you villain?'

There was really very little that the captain could possibly say, so he fell on his knees before Periander. 'Spare me, my lord king. I did not want to kill Arion – it was the sailors who forced me to rob him. Spare me, I beg of you!'

'You deserve no mercy,' said the king sternly. 'Arion, I will allow you to sentence this man and his crew. If you say that they should die, I will order them to be executed at once.'

The captain gave a scream of terror, and the sailors, who were being held outside the throne-room and could hear all that was going on, set up a wailing that could be heard in the palace grounds. 'Spare us!' cried the captain again, turning to Arion. 'Your money and jewels are safe on board ship, and you will have lost nothing!'

Arion, like so many musicians, was a merciful man, and now that he was safe he could not bring himself to take revenge. 'I will spare you,' he said, 'on condition that you return all my prizes, and promise that you will never again try to rob a helpless traveller. I think you should stay in prison for a while, but then you can be set free to sail once more.'

Periander tugged his beard. 'You are very lucky,' he said to the captain, and shook his head. 'I think you should be executed immediately, but I agreed to let Arion decide what is to be done with you, and I will do as he asks. Guards! – take these men off to prison, and keep them there until they have had time to think over their wickedness. Send a party down to the ship to collect Arion's prizes, and take the money and valuables to his home. I hope that after this, Arion, you will stay in Corinth, and not go off on stupid journeys to other lands.'

Arion knew that this was sound advice, and so he stayed at the king's court, where his singing and playing became even better than they had been before. He often thought about the dolphin, and wondered where it was, but he knew that the friendly animal had been right; a life at court would not have suited it, and like all dolphins it was far happier living out its life in the depths of the ocean.

So the whole story ended happily, and much later, when the dolphin died at a great age, the god Apollo decided that it had earned a place in the sky. So the dolphin was brought back to life and set among the stars, where you may still see it on any summer evening, when it came to the rescue of the great singer Arion.

One interesting point about this story is that both Arion and King Periander are thought to have been real people. They lived about

600 BC, and Arion was almost certainly a clever musician. A part of his *Hymns to Poseidon* survives – and Poseidon, of course, was the Greek name for the god we usually know by his Latin name of Neptune.

The constellation of Delphinus, or the Dolphin, is easy to find. It contains no bright stars, but there are several fainter ones so close together in the sky that they make up a conspicuous little group. On a summer evening, find the triangule which is made up of

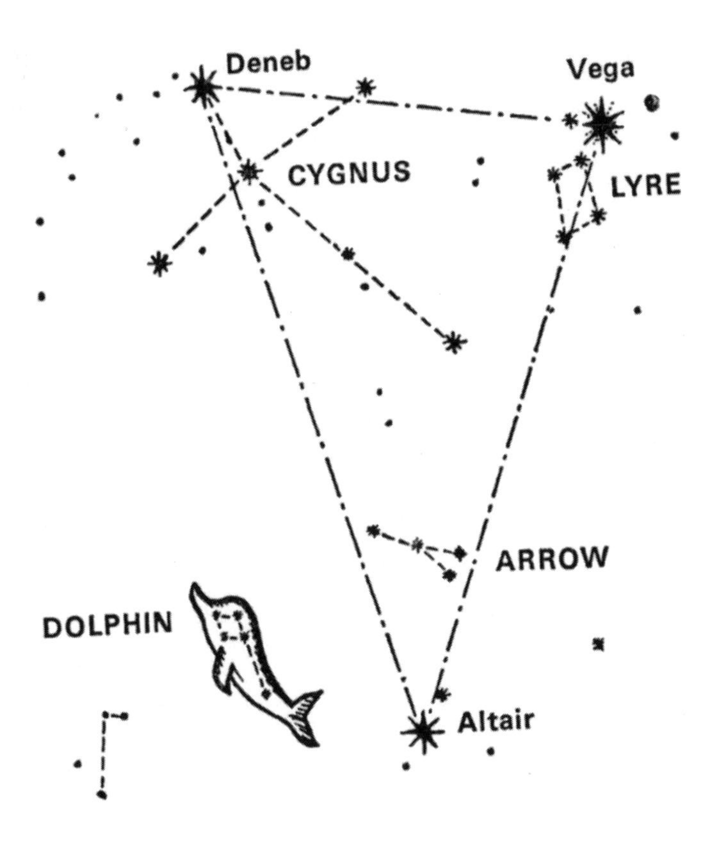

Vega, Deneb and Altair; Vega is extremely brilliant, and you will see it almost directly overhead after the sun has set. Delphinus lies not far from Altair, in the neighbouring constellation of the Eagle. Its leading stars make up the Dolphin's head, so that you will be able to trace the outlines of this brave and friendly creature of the sea.

5

The Three-Headed Dog

You will remember Perseus, who killed the Gorgon and then had a long and happy reign on earth before he was placed in the sky, together with his wife, the beautiful Princess Andromeda. Perseus had several children, and from one of these children was descended the greatest hero of all. This was Hercules, the strongest man who has ever lived.

Hercules was born in the Greek city of Thebes, which was then almost as splendid as Athens itself. Perseus was his ancestor; he was also related to no less a person than Jupiter, the king of the gods, so that from the very beginning it was clear that he was something more than an ordinary boy. It was said that when he was still a small baby, he strangled two poisonous snakes which came near his cot and tried to bite him.

Unfortunately, Jupiter's wife, Juno, disliked Hercules so much that she did everything in her power to hurt him. She tricked

Jupiter into making another prince ruler of Greece, which was not what the king of the gods had meant at all. The other prince, Eurystheus, was Hercules' cousin, but was not the same sort of person. He was neither brave nor wise; in fact he was at best rather a weak, untrustworthy man.

'It is quite wrong that a prince such as Eurystheus should rule over the Greek people,' grumbled Jupiter. 'How much better off they would be with Hercules as king! Why, he is stronger than any lion, and sings more beautifully than any nightingale. As for Eurystheus, he is really nobody. I can't think why you prefer him.'

'You gave me your promise,' said Juno, 'and you must keep it.'

Even though Hercules was not a king, Juno still kept on trying to change him; at last she tricked him into behaving so badly that he was driven out of Thebes altogether. It was not Hercules' fault, but he had to go, and naturally enough he decided to seek advice from the Oracle at Delphi – the most famous Oracle in all Greece.

What he heard made him both angry and sad. The Oracle left no room for doubt. Hercules must go to his cousin, King Eurystheus, and promise to undertake any tasks which might be given to him. There would be twelve of these 'labours' in all, said the Oracle, and Hercules would not be free until he had finished them.

There seemed to be no help for it, so Hercules went off to the king's court. His cousin was not displeased; like Juno, he hated Hercules (mainly, I suspect, because he was jealous of his strength and wisdom), so that he would have been glad to see him killed.

THE THREE-HEADED DOG

'You understand what is to be done?' said Eurystheus, looking down from his throne and wishing that he had half Hercules' courage. 'Twelve tasks, all of them very difficult. If you fail in any one of them, you will never be free, and will by my slave until you die.'

'I shall not fail, cousin. Be sure of that,' Hercules drew himself up to his full height, until he towered over the king. 'The gods are with me, and I shall be able to deal with any problems you give me.'

'We'll see about that,' thought Eurystheus. 'I'll teach this boastful cousin of mine a lesson.' Aloud, he said: 'At least you are young and strong, which is just as well. For the first task, you must go and kill the great black lion that lives in the forests of the country of Nemaea. It has eaten many people, as you doubtless know, and its skin is so tough that no arrow can pierce it. With your great muscles, however, you will certainly overcome the beast.' He smiled, but inwardly he was wondering how long it would take the lion to eat Hercules, too. 'I wish you good luck, cousin.'

Hercules knew quite well what the king was thinking, but he did not stop to quarrel or argue. Instead, he went straight to the wild country where the Nemaean lion lived, found the great animal in its lair, and strangled it with his bare hands, after which he skinned it and made himself a garment which was remarkably strong. It was not long before he was back at the king's court, waiting for his second task.

Eurystheus was surprised to see him, but he was not worried; after all, there were eleven labours left, and surely Hercules would fail in at least one of them? So Eurystheus thought extremely hard, and set his cousin the most difficult tasks he could possibly find.

It would take me a long time to tell you of all Hercules' adventures during the next few years. He slew the Hydra, a nine-headed serpent which lived in a marsh; when he found that he could not kill it by cutting off its heads, since they promptly grew again, he set fire to the undergrowth, and then burned the places on the Hydra's body from which the heads had sprouted. He also killed some remarkable birds which had beaks, wings and claws of iron, and which ate as many men and women as they could find; he captured a bronze-hoofed, golden-horned hind, though he had to chase it for a full twelve months before catching up with it. He cleaned the incredibly filthy stables belonging to King Augeus of Elis by altering the course of a river and sending the waters swirling through all the dirt; and he rid the island of Crete of a fierce wild bull which had been making the people's lives a misery. Another of his tasks was to capture a savage boar which had been roaming the woods of Arcadia, a part of Greece. Hercules seized the boar, but did not kill it; instead he brought it alive to Eurystheus, and the king was so frightened that he ran away and hid in a large jar, refusing to come out until Hercules had taken the boar away.

I must tell you of the task in which Hercules was told to go to the Hesperides, the daughters of the Evening Star, and take the golden apples which grew on the tree in their enchanged garden. Perseus, as you will remember, had visited the garden and had seen the golden apples, but had not tried to take them;

they were guarded by a fierce dragon which never slept. Hercules killed the dragon and took the apples, but when he had shown them to Eurystheus he gave them to the goddess Minerva, who, truth to tell, was keeping a close watch on him. Minerva returned the apples to the daughters of the Evening Star, who were glad to see them back, even though they no longer had a dragon to act as a guard.

Every time Hercules returned, with a fresh task completed, Eurystheus grew more and more annoyed. At last there was only one labour left. However, Eurystheus was a cunning man as well as a cruel one, and he had kept the hardest task until the end.

'You have certainly shown yourself to be clever as well as brave,' he said, looking at Hercules and frowning. 'It may be that the gods have helped you, but for the twelfth task you will have to depend upon yourself alone. You must go to the Underworld, where King Pluto rules, and bring back his three-headed dog, Cerberus.'

This was more than enough to make the strongest hero turn pale, but Hercules showed no sign of fear. 'I have obeyed you up to now, cousin, and I will complete my labours. After that, I shall be free of you.'

'You will have earned your freedom,' said Eurystheus, with a sneer. 'No man has yet entered the Underworld and returned, but – who knows? – you may be the first. We shall see.'

King Pluto, as you will remember, was Jupiter's brother.

He had only once left the Underworld since the time when the gods had first begun to rule; that was when he had carried a girl off to become his queen. The girl, sometimes called Persephone and sometimes Proserpina, was the daughter of Ceres, goddess of the Earth. She was not at all pleased at the idea of marrying a gloomy monarch such as Pluto, but she had very little choice in the matter, even though her mother Ceres insisted that she should be allowed to spend six months of every year under the open skies.

Few people wanted to visit Pluto's kingdom. Generally, only those who had died were allowed to enter – and there was no way out again! Several tunnels were known to lead down to it, and one

Hercules shows King Eurystheus the Arcadian boar

of them was to be found high in the hills; but when Hercules left Eurystheus and made his way across the mountains, he felt more troubled than at any time since he had begun his twelve labours.

'I may as well go into the mountains,' he said to himself. 'At least there is said to be an entrance to the Underworld there, and I must try my luck, even if Pluto finds ways of stopping me from getting in.'

It took him some time to reach the place where he believed the tunnel to be, and for many hours he searched around, wondering whether he had come to the wrong spot. Presently he heard a voice, and turned to find a young man standing beside him. The stranger was tall, and handsome, and merry-looking; the winged sandals and the snake-headed staff showed who he was. As you will

have guessed, he was Mercury, the messenger of the gods, who had once helped Perseus to kill the terrible Gorgon.

'Well, well!' said Mercury, and laughed. 'You seem to be lost, my friend. Can it be that you have missed your way?'

Hercules threw back his head, and laughed in return. 'I have no winged sandals, and so I cannot fly high above the earth and see my path clear,' he said. 'I need your help, as you well know. Tell me, what must I do in order to enter the Underworld? I have wandered over most of the earth, but I have yet to dive beneath it.'

'You will need all your courage,' said Mercury, and sat down upon the ground, folding his arms and rocking to and fro. 'Even I cannot go to the Underworld except when King Pluto invites me – which is not often, I assure you; he and I have nothing to draw us together, and I find him a very dark, gloomy fellow.'

'Is his wife gloomy, too?' asked Hercules.

'Oh, Persephone's not the same sort of person at all,' said Mercury, with a sigh, and stretched himself out, sprawling upon the soft, green grass. 'I must say I'm sorry for her; she has to spend six months of every year in the Underworld, as you know, and I don't think that Pluto can be a very cheerful companion. Besides, there are all kinds of other folk whom I don't like in the least. There are some black witches, called the Furies –'

'Who are they?' said Hercules, as Mercury stopped. Truth to tell, he was not at all anxious to go down into this dark, cheerless place, and he did not mean to stay there a moment longer than was necessary.

'Pluto's servants. He has plenty of others, too,' said Mercury.

'There is Thanatos, for one – he's the god of death, and it's his duty to meet people who have died and who are coming to the Underworld. Well, well, I shall never have to go with him myself. There is something to be said for being immortal!' Mercury gave his friendly chuckle; indeed, it was impossible to picture the gay, fast-moving messenger of the gods as a prisoner in the Underworld. 'I can give you advice, even though I can't come with you.'

Advice was just what Hercules needed. 'Must I go right up to Pluto's palace?' he asked. 'If I can catch his three-headed dog, I can be out in the open air again before Pluto even knows that I have been inside his kingdom.'

Mercury shook his head. 'No, that would be a foolish thing to do. If you get as far as Pluto's throne, I think he will allow you to take Cerberus – if, of course, you promise to bring him back after you have shown him to your cousin. I don't mind telling you that I have spoken to Queen Persephone, and she will do all she can to help you. Listen carefully while I tell you how to reach the Underworld.'

Hercules sat down beside Mercury, and waited.

'The entrance is not very far from here,' began the Messenger of the Gods, and pointed. 'I can take you as far as that. You will go down a long, very dark passage – you'd better take a burning stick with you, by the way, or you won't be able to see anything at all – and presently you will come to a river. This is the River Acheron, and you must make Charon, the boatman, ferry you across.'

Hercules had heard of Charon, who guarded the river which ran round the Underworld. It was Charon's duty to keep a careful watch to see that nobody crossed the Acheron unless he were coming to the Underworld for good; also, Charon had never been known to ferry anyone out again. He was said to be a rude, surly old man, as unpleasant as the river itself.

'I can deal with the ferryman,' said Hercules. 'There are other rivers too, are there not?'

'There are three more, but you will have no trouble in crossing them,' said Mercury. 'There are no poisonous fishes or snakes in them; nothing can live in those dark waters. Then you will cross a glade of dead trees, and at last you will come to the main gates of the Underworld. That is where you will find Cerberus.'

'I don't understand why I'm not allowed to carry him away without bothering Pluto at all,' grumbled Hercules. 'Still, you have been a good friend to me, and I will do as you say. Are you sure that Cerberus has only three heads? I have heard tell that he has as many as fifty, each uglier than the last.'

Mercury stood up, and gave a cheerful laugh. 'That is merely a tale, I promise you. No, Cerberus has only three heads, and I must say that I have never found him quite the terrible beast that he is said to be. Come with me, and I will take you to the tunnel.'

He led the way across the wooded slopes, and after a while Hercules found that they had come to a hole in the ground. It was no ordinary hole; it looked very black, as if it had no bottom, and it was large enough for a man to enter. Hercules noticed that the humming insects avoided it, and that no bee or dragonfly dared to go inside. This, then, must be the tunnel which led to the Underworld.

Mercury pointed. 'There lies your way. As I have told you, I am not allowed to come too, but I wish you good fortune. Remember, you must take great care. If you should make a mistake, then you will never be able to leave the Underworld, and you will have to stay in Pluto's kingdom for ever.'

This was certainly not a nice thought, and Hercules knew that he would be very glad to finish with this twelfth and last labour. Still, there was no sense in delaying, so he said goodbye to Mercury, thanking him again for all his help, and then took a tree-branch, lit it, and stepped into the tunnel. For a few moments he could see clearly, but then the tunnel twisted to one side, and he was plunged into darkness. The burning branch lit up the walls to either side of him, but that was all.

Hercules walked on, hoping that he would come to the river before the burning branch was used up. Every step took him deeper down, and further away from the pleasant sunshine above; the air was damp and chilly, and Hercules shivered, even though he was wearing the skin that he had taken from the Nemaean lion when he had first begun his tasks for Eurystheus.

'This is certainly a dreadful place,' he thought to himself. 'I feel so much alone that I shall be glad to see even the ugly and disagreeable old Charon.'

The branch burned steadily in his hand, but it was almost used up when, at last, Hercules strode round another bend in the tunnel and stopped at the bank of a river. 'This must be the Acheron,' he said aloud, and looked around him. 'Charon! Charon, the ferryman! Where are you?'

'I am here,' came a thick, rasping voice, and a boat glided out of from the shore. 'Who are you, fellow? Unless you have business with King Pluto, I certainly cannot take you across the river.'

'My name is Hercules, and you will ferry me across the water whether you wish to do so or not.' Hercules stepped forward, and gave a frown. 'Pluto does not expect me, it is true, and I do not think that he will be pleased to see me, but he will have no choice in the matter. Take me to the far shore, unless you want me to pick you up and throw you into the river!'

Charon muttered and growled to himself, but he could see that Hercules meant what he said. So with a very bad grace he stepped into the boat, and rowed across the water without saying a word. Nobody had ever spoken to him in such a tone before, and you can understand that he was afraid of this tall, strong man who spoke so bravely about forcing an entrance into Pluto's palace.

'I have no payment for you, but I shall be back when I have carried out my business in the Underworld,' said Hercules sternly, as he stepped ashore. 'Be ready to take me back, boatman, or it will be the worse for you.'

'Your boasts will cause you trouble, mark my words,' hissed Charon. 'You will never come back to this river. Pluto will keep you, and make you his servant.'

'We shall see,' said Hercules, and walked off without bothering to argue. Charon was just as ugly and rude as everyone had said. Hercules was glad to get away from him.

The tunnel was wider now, and Hercules no longer needed the burning branch; there was a dim light, though it was very different from the cheerful glare of the sun. Presently Hercules came to the second river, and then to the third, whose magic waters made anybody who drank them forget all about the world above. Lastly there was the black river called the Styx, which marked the edge of the Underworld. As Mercury had said, the rivers were easy to cross, and there was no need to make use of a boat. Beyond the Styx lay the glade, where dead trees rose up out of the gloom looking like so many dark and terrible giants. And beyond the glade lay a pair of bronze gates, reaching from the rocky floor right up to the roof of the cave. As Hercules came up he could hear a low growling, and when he reached the gates he saw three pairs of eyes gleaming in the half-light.

'Cerberus! Keep back,' called Hercules, and opened the gates, making ready for a fight. 'I have business with your master, King Pluto, and if you try to stop me – well, you will be hurt. Be warned!'

Cerberus reared up on his hind legs. Though he had three heads, he had only one body, so that he looked unlike any other dog either in the Underworld or on the surface of the earth. The heads were of the same shape, but each one barked and growled on its own, so that the cave was filled with noise; as the nearest head poked out towards Hercules, the mouth opened, showing a row of terrible curved teeth.

Hercules stepped forward. 'Keep back,' he said again. 'I have more strength than you, and I am the friend of Mercury, messenger of the gods. I wish you no harm, but you will be very foolish if you attack me.'

Cerberus seemed to understand, and he crouched down, his body low on the ground and all three jaws snarling fiercely. Hercules passed by, closing the bronze gates behind him, and walked on into the heart of the Underworld.

He had expected the place to be dismal, but it was even worse than he had believed it would be. Ghostly figures fled at the sight of him; one or two could be recognised – for instance there was Medusa, the Gorgon who had been killed by Perseus so long before; no longer could she turn men into stone, and she was quite harmless, so that Hercules did not trouble about her. Several times he stopped, trying his best to help poor souls who were in pain or trouble, though there was not really much that he could do in this dark and terrible land.

At last he came to a gloomy building which looked a little like a royal palace. A tall figure stood near, and Hercules saw that this was no pale ghost. In a rather terrifying way the man was extremely handsome, and Hercules knew that it could be none other than King Pluto himself.

'I have heard of you, my boastful friend,' said Pluto in a bitter voice. 'Who else would threaten my ferryman, burst through the gates of my kingdom, and interfere with my subjects as you have been doing? Have a care, Hercules. My powers here are unlimited, and if I wish – well, I can keep you here together with the other spirits who live in the Underworld.'

Hercules took a step forward, and raised his spear. 'That you will never do. You are a ruler of darkness, and I belong to the sunlight. Give me permission to take away your three-headed dog, and I will go in peace.'

Pluto drew his sword. 'You dare to threaten me?' he said, his eyes flashing furiously. 'You must know that I am immortal, and the brother of Jupiter, who rules over all Olympus. I can kill you with this sword, but you will never be able to kill me with your spear.'

'I can try,' shouted Hercules, and thrust with all his strength. Pluto took the force of the blow on his sword, but for a moment he almost lost his balance, and before he could recover Hercules had thrust again. This time the spear caught Pluto in the shoulder, and the king of the Underworld cried out. Then the battle was joined; Hercules, with his tremendous strength, soon found that Pluto was a fighter as skilful as himself, and there is no knowing how it might have ended had not Queen Persephone, Pluto's wife, come out of the palace and called upon the two men to stop.

'You are mad, both of you,' said Persephone, as Hercules and Pluto drew back. By now, both were a little ashamed of themselves; Pluto felt that it was not right for a god to fight against a mortal, even a mortal as strong as Hercules, while Hercules for his part had started to think that he should have been more polite in the first place. Persephone, who was a beautiful young woman who seemed to bring a gleam of sunlight into the depths of the Underworld, stamped her foot angrily. 'Stop it, I say. Tell me the cause of your quarrel, and let me decide between you.'

'I am wounded in the shoulder,' muttered Pluto. 'Wounded, not by a god, but by a common man!'

Persephone wasted no time. She looked carefully at Pluto's injury, and bound it up, though to be honest it was not bad enough to cause any real pain. Then she turned to Hercules, who was standing by with his spear still raised. 'Foolish mortal, I can see that Mercury is the cause of this,' she said. 'But for him, you would never have found your way across the river and through the bronze gates.'

Hercules bowed. 'That is true, my lady. You know, too, why I have come here?'

Persephone smiled, and Hercules felt that she was his friend. 'Yes, I have heard what you wish to do. I think that you should be allowed to try, though I cannot tell whether you will succeed.'

'Carry off my watchdog?' spluttered Pluto. 'Ridiculous. I refuse to allow it. Besides, you would certainly not be strong enough to overcome Cerberus, unless you waited for him to be chained up and then attacked with your spear.'

'I do not want Cerberus to be chained, and I will not use a spear,' said Hercules quietly. 'Listen to me, king of darkenss. You know that I have one labour left; eleven out of the twelve have been completed, and if I can take Cerberus and show him to my cousin, Eurystheus, I shall be free. You would not wish me to remain a slave for the rest of my days?'

Now Pluto, you must understand, was not quite the wicked monarch that so many people believed him to be. He did not like bright lights, or laughter, or green fields, so that the

Underworld suited him very well; but he was not really cruel, and he admired a brave man. So he took Persephone's hand, and looked Hercules full in the face. 'Very well,' he said. 'You can take Cerberus, if you make two promises. First, you must overcome him by your strength alone, without using spears or arrows. Secondly, you must promise that when you have shown him to your cousin, you will take him back to the mouth of the tunnel, and let him return to his rightful place at the gates of the Underworld.'

'I agree,' said Hercules at once. 'I will conquer Cerberus with my bare hands, and I will send him home once he has been taken to Eurystheus' court.'

'You seem very sure of yourself,' said Pluto grimly, 'but if you get the better of Cerberus, you will be the first soul who has done so – and many have tried. Come, then, and see what you can do. I will watch.'

So Hercules and Pluto went back to the bronze gates – leaving Persephone behind; the queen had no wish to see another fight, and she was afraid that Hercules would be badly hurt. As for Pluto, he kept on smiling darkly to himself, wondering whether his three-headed dog would chew this boastful mortal to pieces or break every bone in his body.

Cerberus was waiting, all his three heads turned towards his master. Pluto pointed. 'There he is, Hercules. Take my advice, and go back to Eurystheus alone. I am in a kindly mood, and I will allow you to leave unharmed if you wish.'

'If I go, this creature goes with me,' said Hercules in a firm voice. He laid down his spear, and took off his lionskin, which, as you will

Hercules faces Cerberus

remember, was so tough that no arrow could pierce it. 'Stand back, and watch.'

He crept forward, while Cerberus reared up, hissing and growling. Then Hercules leaped, and landed full on the dog's back. What a yelping and a snarling Cerberus set up! He tried his best to bite Hercules, first with one head and then with another, but he soon found that he was held in a grip of iron. Then he began to struggle, and once he almost threw Hercules off, but before he could close his jaws to bite he was caught once more. For more than an hour the fight went on, until at last Cerberus knew that

he had met his match. He stopped wriggling, and his three mouths hung open, while his tail drooped on the ground.

'Upon my word!' muttered King Pluto. 'I would never have believed it. If I had thought that you had even a chance of conquering my watchdog, I would never have allowed you to try. You cannot really mean to take him to your cousin?'

'That is my purpose, and you have given me your solemn promise.' Hercules stood up, holding Cerberus by the scruff of one of his necks; the dog was really frightened now, and whimpered pitifully, wondering whether Hercules meant to dash him on the hard rocks and make an end of him. 'Send a message to Charon and tell him to be ready with his boat. I will not return to your kingdom, but I will put Cerberus back in the tunnel, as I said I would do. He will certainly know which way to go.'

Pluto was not at all happy about it, but he had no choice, so he had to give Hercules back his spear and lionskin and then let him depart. As Hercules passed through the gates, and walked back across the forest of dead trees, he looked over his shoulder and saw King Pluto still standing there, rubbing his injured shoulder and looking even gloomier than before.

Ghosts were flitting to and fro, and again Hercules caught sight of the Gorgon Medusa, but he did not linger; he was only too anxious to see the last of the Underworld, and he was glad when he had at last crossed the river in Charon's boat and was back in the tunnel that led up to the mountain-top. He had no burning branch to guide him, and he had to make sure that Cerberus did not wriggle free from his grip, but after a while he saw the light ahead, and stepped back into the open air. Never had the sunlight seemed so fresh and glorious. 'I could never live down there,' he thought, gazing back down the tunnel. 'As for you, Cerberus, I must keep my word to your master, but if you have any sense you will stay in these green fields instead of making your way back into the darkness.'

But Cerberus did not like the sun; he growled and squirmed, closing his eyes and longing to be back in the caves he loved. Hercules wished him no harm, and the sooner this business was over the better it would be, so he set off at once for the royal palace.

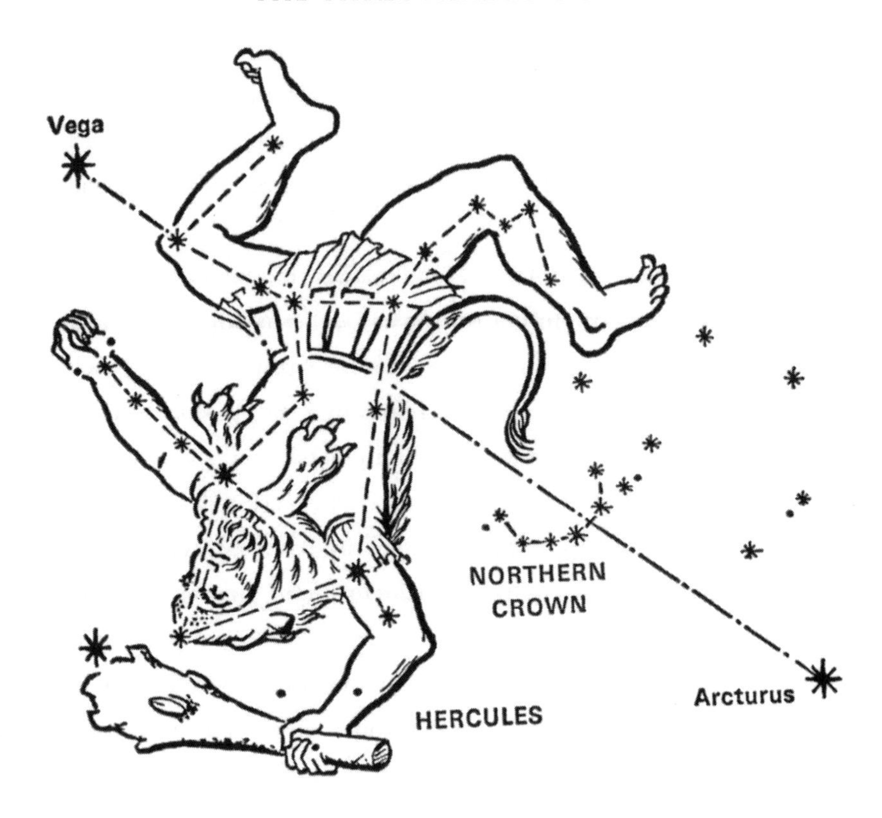

Now and again he met travellers, but none of them would come near. A single glance at the three-headed dog was enough, and they fled, wondering whether this monster had come straight up from the Underworld – as, of course, it had.

When Hercules reached the palace, he did not pause at the entrance. He walked straight through to Eurystheus' throne-room, and flung Cerberus down at the king's feet. 'I have kept my promise, cousin. I have completed my last labour, and I need serve you no more.'

'What …' began Eurystheus, and then suddenly saw Cerberus, who was shaking himself and snarling with all three heads. 'It is not possible! Take it away, Hercules. Take it away, I beg you, before it bites me!'

The king jumped back, but found that he could not escape, and Hercules watched, smiling grimly as Cerberus edged forward. Eurystheus screamed, and shrank back against the wall, pale-faced and shivering. 'Take it away, I tell you. I have given you your freedom – what more do you want from me?'

Hercules gave a deep laugh, and caught Cerberus round the body, pulling the dog back. 'Nothing, my cowardly cousin. You have shown me that you have no courage, and I will stay here no longer. Farewell.'

So saying, he picked Cerberus up and strode out without a backward glance, leaving the king to totter back to his throne and sink down against the soft cushions, thanking all the gods that this dreadful monster had gone. 'Hercules is too powerful,' he said to himself. 'Well, I have done my best to harm him, and I have failed – now I must do no more, in case he takes vengeance on me. I only wish I had never heard of him!'

Yet Hercules had no thought of revenge. He had done all that he had been ordered to do, and it remained only for him to return Cerberus to his master. So he went back to the tunnel in the mountain, and threw Cerberus into the entrance. The dog gave a last whimper, snarled angrily at the blue sky and setting sun, and then turned, making haste back to the inky darkness of the Underworld, which was the only land where he could ever be happy.

'Well done,' said a voice, and Hercules found that Mercury was standing beside him. 'No man could have acted more bravely or more sensibly. I have heard what happened. I only wish I could have seen Pluto's face when you thrust at him with your spear,' he added, breaking into his usual boyish laugh. 'Such a thing has never happened before, and it will be a very long time before it happens again, if indeed it ever does. You have made amends for your misdeeds, Hercules; you can go back to your home in the city of Thebes, and I think you will find that people there have forgiven you for behaving badly so long ago.'

Though Hercules had finished his twelve labours, and though Mercury had been right in saying that the Thebans would welcome him home, he had more adventures in his life. He died at last, but he did not have to go to the Underworld. Jupiter had been keeping a watchful eye on him, and at once the king of the gods sent a thundercloud to take Hercules and carry him up to Olympus. Juno, who had hated him so, changed her mind when she saw him, and not only made him welcome but even allowed him to marry her daughter. So Hercules became an immortal, and it is no surprise that we should be able to see him today shining down from among the stars.

For some reason or other, Hercules is not so conspicuous in the sky as one might have expected him to be. The best way to find him is to start with two really brilliant stars, Vega and Arcturus. Arcturus lies in a line with the curve of the Great Bear's tail, and Vega, shown in the illustration on page 107, may be found by using two more Bear stars as pointers. Arcturus is decidedly orange, while Vega is blue. As you will remember, Vega is almost overheard on summer evenings.

Two groups fill the space between Arcturus and Vega. One of them, Corona Borealis (the Northern Crown), is small but easy to recognise, because its stars make up a conspicuous little semicircle. Hercules is much larger, but his stars are rather faint, and there is no really well-marked pattern. However, if you look hard enough you will be able to see the outline of the great hero who accomplished so many splendid deeds.

Two more constellations are linked with Hercules. One, Leo, is named in honour of the Nemaean lion whose skin Hercules took. To find it, go back to the Great Bear, and use the pointers 'the wrong way', so to speak, so that instead of looking towards the Pole Star we look in the opposite direction. We shall come straight to Leo, which is extremely strong on spring evenings. Its brightest star is Regulus, which makes one of a well-defined curved line of stars known as the Sickle. The rest of Leo is made up of a triangle of stars, one of which (Denebola) is as bright as Polaris.

Leo is high in the south after sunset in the spring. Below it lies Hydra, the great serpent which Hercules killed in the marshes. Hydra is the largest constellation in the sky (not counting the ship *Argo*, about which we shall hear more later, but which astronomers have now split up into several parts). However, Hydra contains only one moderately bright star. This is the reddish Alphard, the 'Solitary One,' which is easy to find because

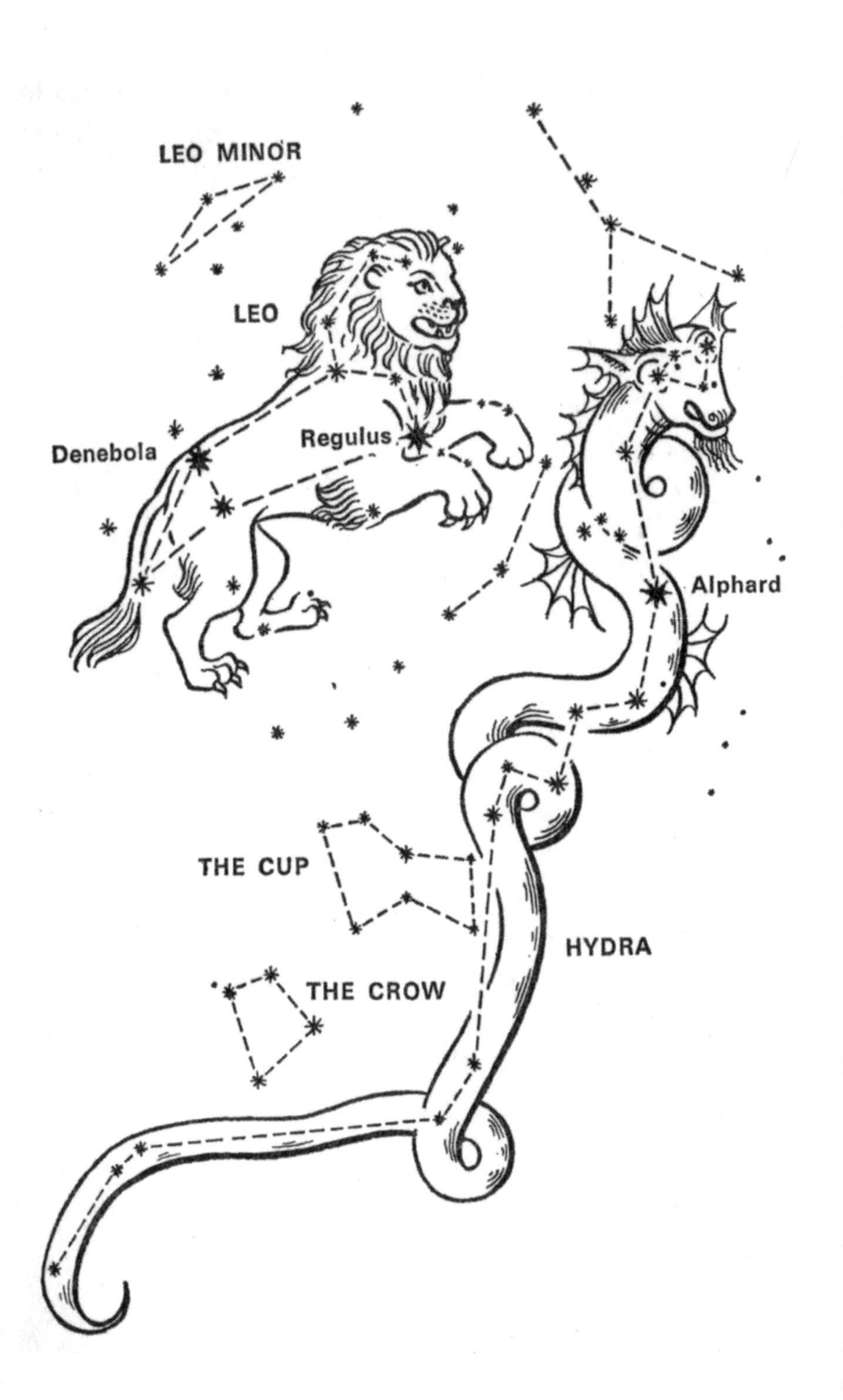
LEO MINOR
LEO
Regulus
Denebola
Alphard
THE CUP
HYDRA
THE CROW

it is so isolated. It is rather strange that the Lion should be so brilliant while Hercules himself is not; but at least you can see Hercules on any summer evening when the sky is clear, and the memory of his triumphs will never fade.

6

The Chariot of the Sun

Jupiter and his queen, Juno, had several children. Most of them were tall, brave and handsome, as you might expect; but there was one son, Vulcan, who was quite different. He was lame, and extremely ugly.

'He's a strange sort of god,' the other Olympians used to say, as Vulcan limped about. 'Oh, he's clever; we all know that, and nobody can handle iron and bronze as well as he can. He looks just like an old blacksmith!'

Vulcan was indeed the blacksmith of the gods, and at last everyone became used to his ugliness. He constructed all manner of iron and metal ornaments, as well as thrones and even gates, so that he certainly made himself useful. One piece of work which made him particularly proud was a golden chariot, which crossed the sky every day carrying Helios, the sun-god.

Helios – we usually call him by this name, though it is Greek and not Latin – was the god who gave the world its light. He saw everything and knew everything, as he could easily do from his position in the sky. Each day he would start off from the swampy land in the east, which we now call Ethiopia, and climb into Vulcan's wonderful golden chariot. Then the nine white, winged horses would he harnessed, and Aurora, the goddess of the dawn, would unlock the golden gates which also had been made by the skilful hands of Vulcan. The sun-god would drive across the sky, among the stars which he made invisible by his brilliant light; at noon he would reach his highest point, and then he would drive down towards the west, sinking to the sea. There he would find a golden ship (yes – also made by Vulcan), and would sail back through the night, out of view from earth, to make ready for the trip next day.

Helios had one private grumble. When the gods divided up the world, at the start of their rule, Helios had been absent, and so he had been forgotten. When he had complained to Jupiter, everyone agreed that he had been unfairly treated, and so he was given the island of Rhodes. The Greeks put up a tremendous statue there in his honour; it was known as the Colossus of Rhodes, and was ninety feet high. It stood for many centuries, and was one of the famous Seven Wonders of the World, though it has long since been destroyed.

One of Helios' children was name Phaethon. However, Phaethon did not live in Olympus; his mother, Clymene, was a mortal, and so Phaethon was only half a god, which did not really count. He was not even sure who his father was, and one day he had a quarrel with another lad, Epaphus, which was to have tragic consequences.

'I don't believe the sun-god is your father at all,' said Epaphus, kicking at a stone in the road. 'You're telling a lot of tall stories. Why, you've never even seen him!'

'I know, but my mother says that I'm the son of Helios, and that's good enough for me,' said Phaethon, clenching his fists. 'If you do keep on talking like this, I'll fight you.'

Epaphus laughed. 'What good would that do? We're not little boys now,' he said mockingly. 'Even if you punched me in the face and knocked me down, it still wouldn't make me believe that you are the son of Helios. Go away and play, Phaethon.'

He walked off, and Phaethon stared after him, angry and hurt. He knew that his father was a god, but how could he make Epaphus and the other boys believe it? It was quite true that he had never seen Helios, except in the sky, and neither had his mother Clymene seen him for many years. At last he went straight to Clymene and asked her about it, which was certainly the most sensible thing to do.

Clymene sighed, and put her hand on Phaethon's shoulder. 'I am not a goddess,' she said sadly, 'but your father is indeed a god. He can spend no time with me here, because he has to drive the sun-chariot every day, and we cannot meet very often. But you can go to him if you wish; he will know who you are, and perhaps he will even take you on one of his journeys across the sky.'

Phaethon looked towards the sun, which was just sinking in the west. 'So my father is up there,' he said. 'It's wonderful, Mother. I'll go, but how shall I find the palace? It must be many miles away.'

'You will find it easily,' said Clymene. 'It shines so brightly that it can be seen across a great distance, and it lies due east, so that you cannot lose your way. I wish I could come with you, but I am growing old, and I do not think that I could make a journey like that. Give your father my love, Phaethon, and tell him to come and see me when he can persuade someone else to drive his chariot for just one day.'

It was a pity that Clymene said this. She meant no harm, but it made Phaethon start to think deeply. Rather than go for a ride in the chariot, how much better it would be to drive the nine winged horses for himself! But for the moment he said no more, and made himself ready for the long journey.

He walked east for many days, and left his home far behind him. He was young and strong, but even so he was starting to become very tired before he reached the marshy land in which the sun-palace lay, and several times he had to stop for sleep. But at last he saw a glow in the distance, and he quickened his step. 'That must be the place,' he thought. 'It is two hours yet before dawn, so my father will still be there. I wonder if he will be surprised to see me?'

As he drew nearer, he saw that the palace was even more splendid than he had expected. The roof was of shining ivory, and the columns were glittering gold; the whole building was so bright that he could not bear to look straight at it. Still he walked on, until at last he passed through the golden doors and entered the great hall. There, on a magnificent throne, sat Helios, the sun-crown on his head sending out brilliant rays that lit up every corner of the hall.

Phaethon stepped forward. 'Father! Do you know me? I am Phaethon, your son, and I have been sent here by my mother Clymene. Are you glad?'

Helios stared in surprise, and then stood up. 'Phaethon! I knew you would come one day. My son!' He put his arms round the young man's shoulders, and embraced him. 'How many times have I wanted to see you! My thoughts are often with you and your dear mother, and I wish with all my heart that I did not have to drive my chariot every day. Come and sit by me, and tell me what has become of my wife and all my old friends.'

Phaethon sat down on the golden throne, and told his father all about his life at home; how Clymene had told him about her marriage with the sun-god, and how sad she was at not seeing him more often. Then he explained how Epaphus had teased him, and had doubted his word. 'I nearly hit him,' he said, 'but that wouldn't have done any good, would it? Besides, I did so much want to see you.'

Helios smiled. 'And I wanted to see you, Phaethon. I shall have to be off in half an hour, because dawn is near, but I shall be back this evening. Of course I shall be able to show your friends that I really am your father. I am not quite sure of the best way,' he added, 'but I will do anything that you ask me.'

Now this was just what Phaethon had been hoping for. 'You really mean that, Father? You will give me anything I want?'

'Of course, if it lies within my power,' said Helios. 'I swear it, my boy. What have you in mind?'

Phaethon sat bolt upright, and drew in his breath sharply. 'I want to drive the sun-chariot,' he said. 'Just once, Father. I know I can do it.'

Helios gave a cry. 'You would never be able to keep the horses in check. They are winged, you know, and are not easy to handle. You would be killed.'

'I wouldn't,' said Phaethon stubbornly. 'I'm a good rider, Father – and you did give me your solemn promise.'

'I – I never dreamed that you would ask such a thing,' muttered Helios. 'If you keep me to my word, Phaethon, I must let you take the chariot, but I beg you to change your mind and ask for something else. The horses have never known a driver other than myself, and you cannot hope to know your way among the stars.'

But Phaethon would not listen; nothing else would satisfy him, and since the sun-god had given his solemn promise there was nothing to be done. With great reluctance, Helios took his son out to the golden chariot, to which the nine white horses had already been harnessed. They made a fine show, rustling their wings and neighing loudly; they were impatient to go, since they all enjoyed their daily flight across the sky. Helios felt more troubled than ever.

'You can see how wild these horses are,' he said. 'If you cannot hold them and guide them, they will plunge earthward, scorching the land as they bring the chariot down. Again I ask you to release me from my promise.'

Phaethon shook his head. 'I know I can drive the chariot, Father. Let me go!'

'I have no choice. I cannot break my word,' said Helios, and gazed upward at the sky; dawn was near, and there were few stars left, so that there was little time to spare. 'Listen, then. Once you have put on the wreath of sunbeams, take great care; remember that they give light to all the earth. The horses ought to know their way, but you must not let them go too near the north pole of the sky, and you must also avoid the coils of the great dragon which circles the polar star. I shall wait here, and, believe me, I shall be more than glad to see you back safe and sound!'

By now Phaethon was starting to feel rather frightened, but he did not want to draw back at the last moment, so he climbed into the chariot and felt the reins. The horses looked round, and whinnied in surprise. Phaethon almost called out 'Very well, Father, I will not go!' but he felt that he would be showing himself up as a coward, and he did so want to be able to go home and boast to Epaphus and the other boys that he, Phaethon, had actually driven the chariot of the sun.

'For the last time, stay here!' said Helios, his voice harsh with anxiety.

But again Phaethon shook his head, and the sun-god stepped back, throwing up his arms in despair. Then Aurora, goddess of the dawn, opened the golden gates that Vulcan had fashioned, and the horses spread their wings. Next moment they were in the air, and Phaethon shouted with excitement. 'This will show Epaphus!' he called out, and waved good-bye. He had a last view of his father standing silently by the gates, and then he was aloft. He was flying!

For a few minutes all went well. The Ethiopian land faded below, and Phaethon drove past the fleecy clouds that had appeared in the blueness of the sky. He could see the stars, though people on earth could not, and he jerked on the reigns.

'Up!' he called. 'Up, my fine horses. Climb into the heavens!'

Then, suddenly, he knew that something was wrong. The horses were not flying as steadily as usual; they could tell that the firm hand of Helios was not there, and they began to plunge this way and that. Phaethon's fears came back to him, and again he cried out, but this time in terror. 'Stop!' he shoulted. 'Don't fly so fast, or we shall be across the sky too soon. Slow down!'

He gave the reins a violent tug, but it was too late. The winged horses had taken fright, and they bolted, their wings flapping to and fro with furious speed. Instead of flying upward, the chariot swooped down, and although he did his best Phaethon could not stop it.

Again he called. 'Father! I'm sorry. Come quickly!' but it was no use; Helios was far below, and in any case it is never possible to turn back the chariot of the sun. Down, down went the horses, and the rays from Phaethon's wreath of sunbeams caught the clouds, setting them on fire and making them glow an angry red. Now they were below the clouds, and the very earth started to burn. Trees blazed, and a wave of fire swept across the Ethiopian land. The people screamed in terror as the heat scorched them; their skins turned black – and to this day the Ethiopians have black skins.

Phaethon screamed more loudly than any of the earth-dwellers, and he tried to force the horses upward. But by now the winged steeds were as terrified as Phaethon himself, and they swept low over the ground. The fair green land began to smoke, and became wild desert, still known to us as the Desert of Libya; cracks opened in the ground, and everything was blackened and ruined. The cracks were so deep that they reached as far as the Underworld, and King Pluto, far below in his gloomy kingdom, gave a shout of anger and alarm. 'What is this?' he called in a thunderous voice. 'Jupiter! Jupiter, my brother, what is happening above? Light is entering my land, and I cannot bear it. Close those cracks in the earth before any more of this terrible sunshine can come in!'

Jupiter, sitting far above in Olympus, heard Pluto call, and he looked down. 'This is madness,' he said furiously, and turned to Mercury, the messenger of the gods, who was usually at his side. 'Go to the land of Ethiopia, and find out why the sun-god is not in his usual place. Something is badly wrong, and if it continues I very much fear that the whole earth will be destroyed. Make haste!'

So Mercury flew earthward as fast as his winged sandals could carry him. It did not take him long to get the whole story from Helios, who was more worried than ever, and before many minutes had passed he was back in Olympus, telling Jupiter what he had found out. 'I do not wish to harm the boy,' muttered the king of the gods, 'but what else is there to be done? Mercury, perhaps you can go to his help –'

Mercury shook his head. 'I would help him if I could, but neither I, nor you, nor anyone else – man or god – can hope to go near the sun-chariot by day without being burned to ashes. We must wait

The deadly thunderbolt hurls Phaeton from the chariot

and see whether this foolish Phaethon can steer the horses back to their true path.'

Meanwhile, Phaethon had managed to make the chariot fly higher once more; he was back among the stars, but he was not following the usual course of the sun. Instead of travelling southward, he could not stop the horses from bolting northward, towards the pole of the sky which Helios had warned him to avoid. Suddenly he found that he was near the terrible dragon, whose huge body spread out across the heavens. The horses saw the dragon, too, and they reared and neighed. Phaethon's hands slipped from the reins, and he sank down in the chariot, the wreath of sunbeams pouring out its radiance on to the dragon's form.

Once more the horses turned downward, and the ground smoked as the fields turned from green to black, while whole towns crumbled into ruin. Phaethon screamed, but it was no use; King Pluto was shouting too, and now Helios climbed up into Olympus, where Jupiter was waiting. 'My son!' he called, half-sobbing. 'My son, why did you do it? Mighty Jupiter, save him by your magic powers!'

'I cannot save him,' said Jupiter, and pointed. 'Even I, ruler of the gods though I am, cannot go near the chariot. If I let Phaethon fly on, the whole earth will be burned up. If I hurl a thunderbolt and strike him down, then the horses will go back to their stable. I can do nothing else!'

'Stop!' pleaded Helios. 'Phaethon is my son. You cannot strike him dead –'

Jupiter paused. 'What else is there to be done?' he said angrily. 'Look. The earth is scorched now, and I can see that the chariot is coming close to the Scorpion of the sky which once caused the death of Orion. Is there anything I can do except strike Phaethon with a thunderbolt?'

'I – I do not know,' said Helios in a low voice. 'Wait, I beg you. If he passes the Scorpion, he may be safe yet.'

But the chariot was close to the horrible creature, and as the horses caught sight of it they plunged aside once more, turning downward towards the earth. Phaethon was too frightened even to try to grab the reins, and far below he could see the smoke rising from the desert. 'Father!' he cried again. 'Father, save me!'

Helios could not reply; neither could Mercury. Then Jupiter

raised himself up in his throne, and aimed one of the deadly thunderbolts that he used when danger was close. 'I do not want to do this deed,' he said gravely, 'but it is the only way. Do not look, Helios. Turn away until it is over.'

He waited for a second more, and then hurled the thunderbolt. It shot across the sky, and caught Phaethon full in the body; the shock threw him out of the chariot, and he fell – down, down, down, until at last he plunged into the broad river Eridanus, which flows through the boot-shaped land of Italy. At once the horses became calmer. They had no driver now, and their flight became less wild, until at last they turned and came back to the marshes of Ethiopia. The wreath of sunbeams had been quenched in the river, and darkness fell upon the earth, broken only by the red glow of the fires that had been started during Phaethon's headlong flight. The day was over, even though it had lasted only a few hours.

'I did not want to do it,' said Jupiter at last. 'With all my heart I am sorry.'

'I was to blame,' whisphered Helios. 'I have learned my lesson. Never again will I allow another man to take my place, even if it means breaking every promise that I have made during my whole life.'

I am afraid that this is not a very cheerful story. We must all feel very sorry for Phaethon; after all, he had meant no harm, though it was foolish of him to take advantage of the sun-god's hasty promise.

Phaethon himself is not to be found in the sky, but there are several things to remind us of his ride. First, there is the shining

The Milky Way

band which we call the Milky Way, said by the Greeks to be due to the wreckage left by the chariot; it extends in a broad belt right round the sky, and if you have a telescope you will be able to see that it is made up of large numbers of faint stars.

Modern astronomers can explain it easily enough. The stars in our system, or galaxy, are arranged in a flattened shape, as shown in the picture, so that when we look along the thickness of the system – that is to say, from right to left or from left to right – we see a great many stars almost one behind the other. It is this which causes the Milky Way effect. We now know that the stars in the Milky Way are not really packed close together, even though they look so crowded, but there are so many of them that you could not hope to count them; it is thought that the galaxy contains at least 100,000 million stars, and each is a sun.

The Milky Way runs through famous constellations such as the Twins, the Eagle and the Swan, as well as Perseus and Cassiopeia. On a dark, clear night it is very conspicuous and beautiful, though people in cities never see it properly.

As for the Dragon which so frightened the nine winged horses, you can find it not far from the Great Bear and the Pole Star, sprawling across the sky and reaching almost as far as the brilliant Vega. There are no bright stars in the Dragon, but if you look carefully you will be able to find it without difficulty – provided that the sky is dark enough for you to see faint stars.

You can also find Eridanus, the river into which Phaethon fell. It is said to be the same as the Italian river now called the Po. In the sky, Eridanus begins close to Orion, not far from the lovely white star Rigel; it is not really easy to identify, because its brightest star, Achernar, is too far south to be seen from England, but it winds its way down to the horizon in a long line of stars about as bright as those of the Dragon. It is best seen in late autumn and winter.

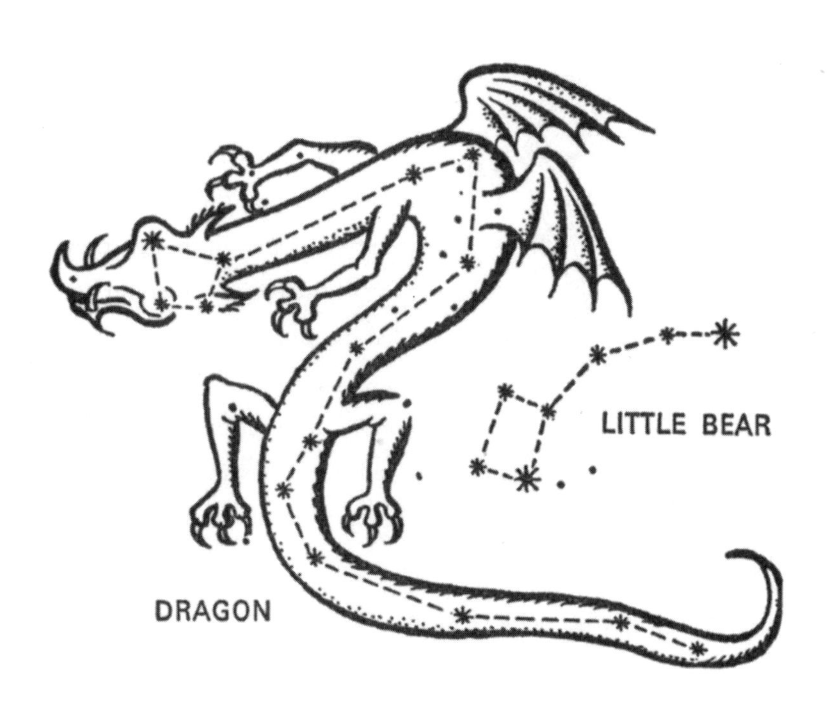

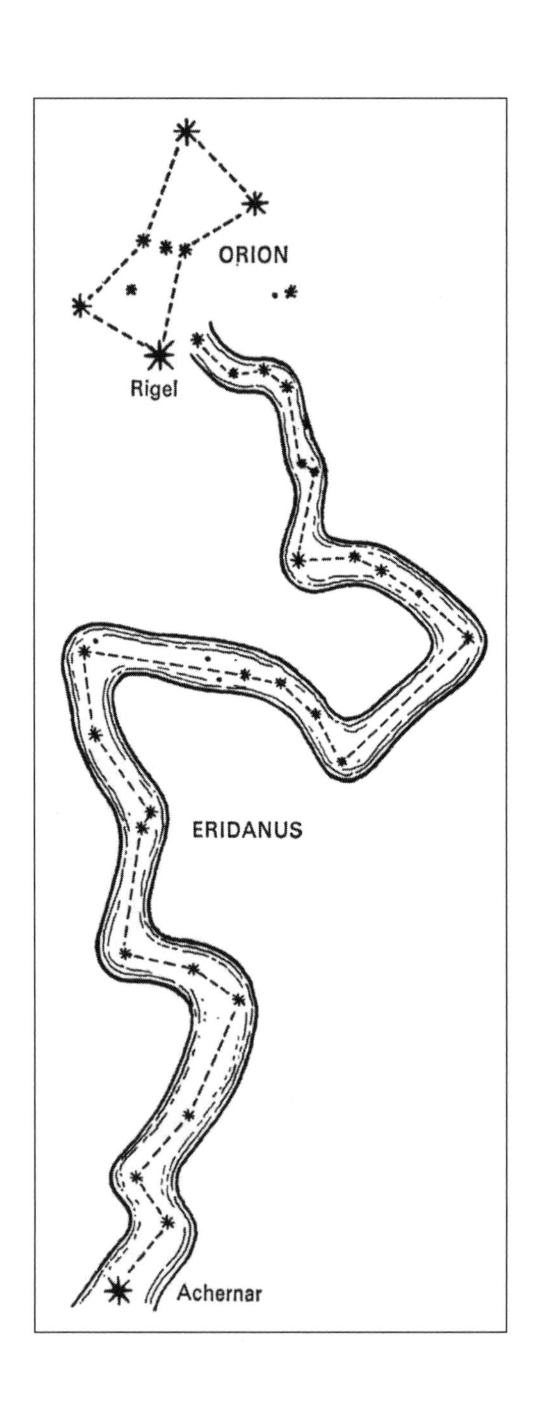
ORION
Rigel
ERIDANUS
Achernar

Phaethon was to blame not because he was wicked, but because he was stubborn and boastful; after all, who but the sun-god could hope to give the Earth its light? And since that dreadful day, nobody but Helios has ever dared to drive the Chariot of the Sun.

7

The Northern Bears

Juno, the queen of the gods, was not always kindly by nature. As her husband was no less a person than Jupiter, the ruler of both earth and sky, one might have thought that Juno would have been well satisfied; and in a way, she was happy enough. Unfortunately she was always jealous of other people, and she liked to think of herself as more beautiful than any other god or mortal.

Juno was certainly very lovely indeed. Even Venus, the goddess of love, and Minerva, the goddess of wisdom, were no more beautiful than Juno, and in the ordinary way all three immortals were sensible enough not to discuss the matter (Minerva, in any case, was much too clever to start an argument which could lead nowhere, and which might cause a serious quarrel). But there were lovely women on Earth too, and one of them was Callisto, daughter of King Lycaon of Arcadia.

Arcadia is a part of Greece. Usually there were many woods and forests there, but at the time when Callisto grew up the whole country was still scorched and blackened from the effects of Phaethon's tragic ride in the chariot of the sun. Jupiter, who had saved the world by hurling the thunderbolt which had toppled Phaethon into the waters, did all he could to put matters right, but even he could not make the trees and grasses grow quickly, and it was some years before Arcadia was green and pleasant once more.

'I should have acted more quickly,' grumbled the king of the gods, looking down from his palace high above the clouds. 'It has been a real lesson to me. Still, it can never happen again, which is just as well,'

At last the forests had grown once more, and the wild animals had come back. Meanwhile, Juno had been watching the beautiful young girl who was King Lycaon's daughter. Callisto was more than ordinarly lovely, and Juno began to feel rather jealous of her. Callisto, of course, did not know that she was being watched; she was not interested in court life, and much preferred being out in the woods, taking part in hunting. King Lycaon allowed her to go her own way, no doubt thinking that she would change her habits when she became older.

'Have you noticed that princess?' Jupiter said once, when Juno was sitting beside him in the palace of Olympus. 'She seems remarkably good-looking, and she is a fine hunter too. I really wonder that she doesn't start to think of getting married.'

'Beautiful? Would you say so?' said Juno, looking sideways at him. 'She is not ugly, I agree, but you couldn't class her with Venus, Minerva, or with me.'

Jupiter rubbed his forehead. 'I'm not so sure,' he said, thoughtfully. 'Dress her like a goddess, and I believe she would be the equal of any of you ...' He broke off suddenly. 'No, of course not. You're perfectly right; she is nothing more than an ordinary Greek princess.'

The truth of the matter was that Jupiter knew all about his wife's jealousy. He was deeply in love with her, but he was not blind to her faults, and he realised that his idle talk might easily put Callisto in danger. Juno was quite capable of harming her if she felt so inclined, and Jupiter had no wish for anything of the sort to happen. Unfortunately, the damage had been done. Juno felt that she had been insulted, and that her husband really preferred the girlish beauty of Callisto to that of his own wife. She said no more, but next day, when Callisto was out hunting as usual, it became clear that something was wrong. Her shots missed their targets, and several times she only just escaped being badly wounded.

Jupiter was worried. He guessed that Juno was the cause of the trouble, but what could he do? It would have been useless to ask his wife to forget what had been said, and in any case Juno would not believe him if he swore that he did not really think Callisto to be as beautiful as a goddess. Neither could he watch the girl all day, because he had much too much to do. He certainly did not want to kill her, and neither could he very well snatch her up into

Olympus, since Juno would make all manner of mischief. At last he decided that he would have to speak.

'I am sorry you misunderstood what I had to say about that Arcadian princess,' he began, hoping that he sounded more sincere than he felt. 'Really, my dear, I feel that you are making a great deal of trouble about nothing. I can see that you are planning to injure her …'

'So you have been spending your time in watching her?' said Juno, with a rather unpleasant smile. 'Just as I thought. Well, you need not trouble. I certainly wouldn't bother to harm her.'

'She is a good shot with a bow and arrow,' muttered Jupiter, 'and yet she has not hit a single animal all day long. If she has really lost her skill, she will be torn to pieces by some creature before long.'

'Would that upset you dreadfully? I thought you said that you weren't interested in her,' said Juno, and turned away. 'I really can't waste any more time talking about her.'

Jupiter pretended to be satisfied, and as the weeks went by he hoped that the trouble had been forgotten. Meanwhile, Callisto began to hunt less and less; as her father had hoped, she fell in love and married. King Lycaon and his subjects were well pleased, and the wedding celebrations went on for three days. Jupiter, wisely, paid no attention, and neither did he take any notice when after a year or so Callisto had a son, a boy who was named Arcas. He only trusted that Juno would take no notice either.

All might have been well but for a chance remark made by Minerva, who, as goddess of wisdom, was not in the least jealous of Callisto or anybody else. 'What a lovely woman that princess has become,' she said. 'She was beautiful enough even as a girl, but now that she has grown up – well, she really outshines us all. She has a sweet nature, too.'

Jupiter looked uncomfortably at Juno, and saw that she had drawn in her breath sharply, while her expression was the very reverse of sweet-natured. It was unlucky that several other goddesses joined in, all praising Callisto's beauty, and when Juno swept away in anger it was plain that there was bound to be serious trouble. Jupiter did not know what was the best thing to do, so he simply waited.

Next day Callisto went into the woods which lay around her father's royal palace. She was not hunting; now that she had her baby son Arcas, as well as her husband, she seldom went very far away, but she still loved the forests and glades. Suddenly she stopped, and looked up in surprise. Ahead of her stood a woman whose beauty matched her own, but whose expression was stern and threatening.

'Who are you?' stammered Callisto, rather frightened.

'Do you not know the queen of Olympus? I am Juno, wife of almighty Jupiter.' Callisto shrank back, more frightened than before. 'I have heard that you believe yourself to be lovelier than any goddess. You have been telling lies about me!'

'Lies? I – I do not understand,' said Callisto, her voice shaking with fear. 'People tell me that I am beautiful, but I cannot tell whether they are right or wrong, and to be honest I do not greatly care.'

'Liar!' repeated Juno, and raised her hand. 'I do not like liars, as you shall learn.'

Poor Callisto was too frightened to reply, and stood still, wondering what this terrible goddess was going to do to her. She did not have to wait for long. For a moment she was dazzled by a brilliant light, and then, looking down, she saw that her dainty hands had changed; they had become large, furry paws, and she seemed to be covered with hair. She tried to cry out, but instead of words she could only growl. She was no longer a beautiful woman; she had become a brown bear, and she dropped down on all fours, doing her best to plead for mercy.

Juno laughed. 'This is what becomes of being too proud,' she said in a mocking voice. 'You are not so beautiful now, my fair lady. A bear you have become, and a bear you will stay. Be off into the woods, and hide yourself!'

Callisto had to obey, leaving Juno standing there still laughing in that heartless fashion. And the queen of Olympus had certainly behaved in a very cruel manner; she was not usually so wicked or spiteful, and she may have been ashamed later of what she had done, but it was too late, since even Jupiter could not turn Callisto back into a woman. As for the princess' husband, he was beside himself with grief when he found that his lovely wife had disappeared, and so was old King Lycaon, who had no idea of what could have happened to her. It seemed that Callisto must have wandered too far without her bow and arrow, and had been gobbled up by some fierce animal. Arcas, of course, was too young to understand, but as the years passed by grew up into a tall, handsome lad. He often heard about his beautiful mother, and wished very much that he could remember her. Like Callisto, he became fond of hunting, and he too often went out in search of wild animals, though every time he left the palace his father was anxious until he could see him on the way home. 'My dear wife vanished without trace,' he used to say. 'Do not go far, Arcas. You are all I have left now.'

Arcas would laugh in reply. 'I can look after myself, Father. Don't worry.'

One day Arcas was out in the forest, some miles from the palace. He had several arrows left, but was wondering whether it was time to return when he heard a rustling sound. There, within a few yards of him, stood a huge brown bear, growling almost as though it were trying to speak. Arcas gave a shout, and pulled up his bow, fixing an arrow and taking careful aim.

You will have guessed that the bear was none other than Callisto. She had lived in the woods ever since that dreadful day when she had met Juno, and she had often seen Arcas; now, at last, she had made up her mind that she must try to make him understand what had happened. Unfortunately Arcas could not have the slightest idea that he was facing him mother, and he drew back his arm, ready to release the arrow that would certainly end Callisto's life.

But Jupiter was watching. High above, in Olympus, he had kept a careful eye on both Callisto and Arcas, and now he cast one of his most powerful spells. He was only just in time. In another second the arrow would have flown, but before Arcas could release it he, too, had been turned into a bear. With a mighty stride Jupiter jumped down from Olumpus, seized the two bears by their tails, and swung them up into the sky, setting them among the stars for ever.

So the story had a happy ending, after all, and both the princess and her boy can be seen shining down on any clear night of the year – Callisto as the Great Bear, and Arcas, her son, as the Little Bear, which marks the north pole of the heavens.

The legend of the Bears ends by telling how Jupiter, in swinging the two animals into the sky, stretched their tails, which are still remarkably long. However, we have to admit that it takes a good deal of imagination to make a bear out of either constellation. The Great Bear is more generally known as the Plough, though in fact the seven bright Plough-stars make up only part of the Bear group. The Little Bear is of rather the same shape, but it is much fainter. Americans call them the Big and Little Dippers.

The Plough is probably the most famous of all constellations, though its stars are not so brilliant as those of Orion. It never sets over England, and so can always be seen whenever the sky is dark and clear. Once recognised, it will never be forgotten. It may be overhead, as during spring evenings, or else in the northern part of the sky. Look at the second star of the 'handle,' Mizar, and you may glimpse a much fainter star (Alcor) close by. If you can see Alcor without a telescope, you may be sure that there is no mist or thin cloud about.

The two stars at the far end of the Plough are called the Pointers, because they show the way to the Pole Star, Polaris. Polaris is the leader of the Little Bear, and marks the north pole of the sky, so that it seems to stay almost still while the other stars move round it – completing one full circle in twenty-four hours. This has nothing directly to do with Polaris itself; the Earth's axis of rotation simply happens to point towards it, as shown in the illustration.

The rest of the stars in the Little Bear are fainter, apart from the rather reddish-coloured Kocab, which is about as bright as Polaris

Jupiter swings the two bears into the sky

and is often called 'the Guardian of the Pole.' However, it is quite easy to recognise the outline of the constellation, provided that the sky is sufficiently clear and perfectly dark.

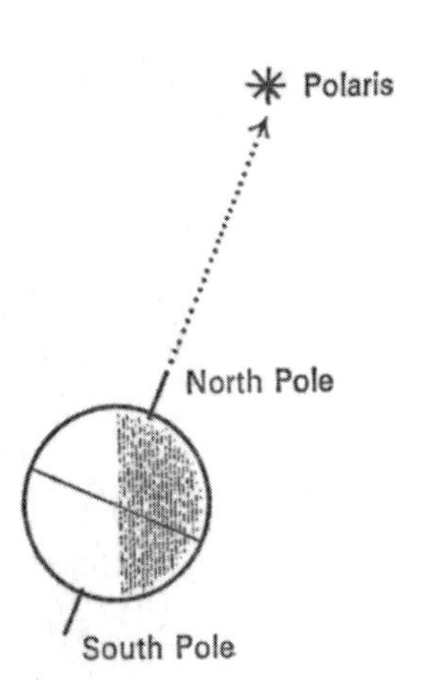

The two Bears are well known to anyone who has the slightest interest in astronomy. They are always there, shining down in their friendly way to remind us of the story of the beautiful Princess Callisto and her handsome son.

8

The Voyage of the *Argo*

Upon Mount Pelion, a grass-covered peak overlooking the sea, there lived a most remarkable person. His name was Chiron, and he was a centaur; from the head to the waist he had the form of a man – and a handsome, clever-looking man at that – but below he was a horse. As he clattered about on his hoofs, swishing his tail, he must have looked very strange, and yet he was not in the least frightening. For Chiron was a teacher, and as noble as he was wise. Many of the Greek heroes had come to him, and among his pupils were Hercules, Orpheus the musician, and even Æsculapius the great doctor.

One day Chiron was standing beside the entrance to his cave, staring out across the waters and thinking of nothing in particular, when he saw a small boy coming up the mountain path. Chrion was not in the least surprised; he was used to visitors, and was always glad to see them. He tapped the ground with his front hoof, and called out, in his deep and friendly voice: 'Who are you?'

The boy looked a little nervous, which was easy to understand. 'My – my name is Jason,' he stammered. 'I have been sent here to look for Chiron, who is said to be a teacher. My father told me that Chiron is half-horse and half-man –'

Chiron laughed, and at once the boy felt more comfortable.

'Your father was quite right, Jason. I and my people are centaurs; as you can see, we have the advantage of four legs, so that we can run more swiftly than the wind. Come and sit down.' He waited until Jason had settled himself, and then went on: 'You are very welcome, but I do not yet know why you have come to seek me. Your father must know of me, but I do not recognise you.'

Jason looked up, doing his best to keep his voice steady. 'I am the son of King Æson of Iolcos, so I am a royal prince.'

Chiron started, and swished the air with his tail. 'Æson, eh!' he muttered. 'Yes, I know King Æson, and a better monarch never lived, but I have heard tell that he has been turned off the throne by his brother Pelias. Is this true?'

'It is true,' said Jason, his mouth quivering. 'My father is a prisoner in his own palace, while my uncle has taken the throne for himself. He says that he is the proper king, and most of the people believe my father to be dead.'

'I see,' said Chiron slowly. 'Yes, I see. So King Æson thinks that Pelias will try to kill you, and you have been sent to me for safety?'

Jason nodded, unable to speak.

'Well,' said Chiron at last, 'you will be safe here, and I will do my best to teach you, as I have taught so many others in the past. Don't look so sad,' he added, breaking into a laugh that made Jason smile in spite of his unhappiness. 'You won't be at all lonely, because a great many people come here, and in any case you are sure to have some fellow-pupils before long. Besides, I can see into the future – not very clearly, I admit – and I know that one day you will be crowned as king in your country.'

So Jason came to live in Chiron's cave, and as the years passed by he grew up into a strong, good-looking youth. As Chiron had said, he was never lonely; there were always plenty of boys of his own age who had come to the centaur's cave to be taught, and Chiron himself was the best of companions. He was certainly well-equipped to teach his pupils how to ride, and Jason covered many miles sitting on Chiron's back as they raced among the mountains. He learned how to shoot, how to hunt and how to find his way by the stars; he learned, too, how to read and write, and how to play upon the harp. Jason proved to be particularly good at doctoring, and Chiron said that he was the cleverest healer who had ever been taught in the cave, apart from the great Æsculapius.

In the evenings Chiron and his pupils would often sit out in the light of the setting sun, singing and making music. Sometimes Chiron would tell them stories about far-off lands, and one evening he was asked about the Golden Fleece. I really forget who asked the question; it may have been Jason, or Orpheus the musician, or Hercules. In any case, Chiron settled himself down and began to speak.

'The Golden Fleece,' he said thoughtfully. 'Yes, that is a famous tale, and I know that it is true. Have you ever heard of the kingdom of Boeotia?'

The boys shook their heads.

'I should have taught you more geography,' said Chiron, and gave a chuckle. 'No matter. Well, it is a land in a distant part of Greece, and some years ago it was ruled by King Athamas. Athamas was a good monarch, and he was very happy with his wife and his two children, his son Phryxus and his daughter Helle. Then, unhappily, the queen died, and Athamas married a wicked woman named Ino.'

'Why was she wicked?' asked Jason curiously. 'What did she do?'

Chiron sighed. 'If I knew the cause of wickedness, I should be far wiser than I am. All I can say is that Ino hated her two stepchildren, and plotted to kill them. She would have succeeded, too, if the whole affair had not come to the ears of Jupiter himself. I admit that I had something to do with it; I could not stand back and see two children savagely murdered.'

'What did you do?' asked one of the boys.

'That does not matter,' said Chiron gently. 'Well, Jupiter called Mercury, the messenger of the gods, and told him that somehow or other he had to save Phryxus and Helle before the queen had time to kill them. Mercury sent a ram – which, as you know, is a large animal – and told the children to climb upon its back. This ram, of course, was no ordinary creature. It could fly, and its fleece was made of solid gold.'

He paused, and the boys waited breathlessly.

'Everything seemed to go well,' went on Chiron. 'The children mounted, and the ram flew off, heading for the country of King Æetes of Colchis, who was no friend of Ino and who would certainly take Phryxus and Helle into his care. This was where I come to the saddest part of my story. As they flew across the sea, Helle, the little girl, lost her balance and fell off. Down, down she dropped, until she landed in the waters and was drowned. That is why that part of the sea is still called the Hellespont, in her memory.'

The boys looked at each other. 'Poor Helle,' said one of them. 'What about Phryxus? Did he fall off, too?'

'No; he was stronger, and he managed to stay on the ram's back until they reached Colchis. As Jupiter had foreseen, King Æetes received Phryxus kindly, and brought him up in the royal palace. When he had grown up, he married the king's daughter, and lived a happy and peaceful life. He died some time ago, but King Æetes still rules. As for the ram, it too died at last, and its golden fleece was placed in a tree in a sacred grove, guarded by a fierce dragon.'

'I think the ram should have been put into the sky,' said Jason.

Chiron smiled. 'That was done, I assure you. If you look up into the stars, you will see the ram there, shining down happily. It is a strange story, is it not?'

'I should say it is!' said another of the boys. 'Surely people have tried to steal the Golden Fleece, though? It must be very valuable.'

'It is certainly valuable, but it would be very hard to steal,' said Chiron dryly. 'King Æetes is very proud of it, and he has made sure that anyone who tries to run away with it will be gobbled up by the dragon before he can do more than set foot in the sacred grove.

Even I would not like to go near, though I have had to fight against more than one fire-breathing dragon in my time.'

Jason and his friends often thought about Phryxus and Helle and the Golden Fleece, but it was only one of the many tales that Chiron told them, and there was always a great deal to learn and do. But at last the boys started to leave to seek their fortunes in the outer world; of course fresh pupils arrived, but by now Jason was a young man, and he felt that it was time for him to depart, sorry though he would be to leave his friend and teacher.

'I can stay no longer,' he said one evening, when he and the centaur were alone on the mountain-top. 'You know how little I want to leave – this is my home, and I have had so many years of happiness here that I can never thank you enough, but I am no longer a boy.'

Chiron smiled, a little sadly. 'I have known what was in your mind, Jason,' he said gently, 'and I am sure that you are right, though I do not want you to go. You have learned as much from me as I can teach, and the time has come for you to go back to your own country and claim the throne for your father. He will be glad to see you.'

'He is still alive, then?' said Jason. He had always wondered, but somehow he had never been able to bring himself to ask a direct question.

'King Æson is very old, but he still lives. Nobody calls him 'king,' I fear; your uncle Pelias has sat upon the throne ever since I first saw you coming up the mountain-path. Pelias is a cruel, heartless man, as you know,' Chiron sighed. 'I have some fresh clothes and sandals ready for you, Jason. Do not delay; if you wait any longer it will only make us both feel more upset when you go – but remember, I am always here, and there is always a home for you in my cave whenever you need it. Climb upon my back, and we will go for a last ride together.'

So they rode back to the cave, and Jason dressed himself in the clothes which Chiron had set aside for him. You may suppose that it was a sad moment for both of them when the time for parting came, and as Jason walked down the path he kept on looking back to the figure of the centaur outlined against the blueness of the sky; but at last Chiron gave a final wave, and Jason passed out of sight into the forest that covered the lower slopes of the mountain.

He walked on for many hours, until after the sun had set. He lay down and rested for a while, but he was on his way again before daybreak, and by early morning he had reached the plain that surrounded the kingdom of Iolcos. It was then that he came to a fast-flowing river, and he hesitated, wondering whether it would be too deep for him to wade across. He could see plenty of rocks in midstream, and it seemed a pity to waste them in walking along the bank in search of a bridge, so he prepared to step into the water. Then, suddenly he heard a voice.

'Young man,' it said, 'I need your help. I am an old woman, as you can see, and I am not strong. Do me the honour of carrying me across the river.'

Jason turned, and found himself facing a little woman who certainly appeared to be very old indeed. 'Where in the world have you come from?' he asked in surprise. 'I did not think that there could be anyone within miles of me.'

'Never mind where I came from,' croaked the old woman. 'What matters is – will you carry me across the stream, or must I try to struggle across on my own weak legs?'

'Of course I will carry you,' said Jason at once, 'and I will try to see that you do not get wet. Climb on my back, then, and hold on as tightly as you can.'

'You will not regret your kindness,' said the old woman, and threw her arms round Jason's shoulders, heaving herself up until he feet were clear of the ground. 'Take care, young man. If you slip, we shall both be soaked.'

Jason tested the ground at the water's edge, and then stepped into the river. The water was cold and fast running, and once he almost fell, but the stream was not really deep and it took him only a few minutes to cross it. Only when he had come to the other side did he realise that he had lost one of his sandals, and he gave a cry of annoyance. 'Look. One of my sandals has been swept away, and I shall never be able to find it. Did you happen to notice where it went?'

The old woman let herself down off Jason's back. 'You need not be sorry about your sandal,' she said, and somehow her voice had changed; it was no longer a croak, but more like the voice of a goddess. 'In fact, you have been very fortunate. It shows me that you are indeed the one-sandalled man of whom the Oracle spoke.'

Jason turned back to the river. 'I wish –' he began, and then stopped. The old woman had vanished; one moment she had

Jason helps the strange old woman across the river

been there, the next she had disappeared as completely as though the earth had swallowed her up. Jason looked around him in astonishment.

'That's strange,' he muttered. 'Who was she, I wonder, and why did she think me lucky in losing one of my sandals? Well, I suppose

I shall find out one day, but there's no point in my staying here any longer.'

Again he set off, thankful that he had reached the level plain and that he would not have to cross any rough mountain country with one of his feet bare. Before long he started to come to some small villages, and he noticed that the people stared at him very curiously; then he reached larger towns, and before nightfall he had come to the chief city of Iolcos, with its fine buildings and its royal palace reaching up into the sky. More and more people crowded out of their homes, and more than once Jason thought that he caught the words 'One sandal! It's the man with one sandal!' but he did not stop to inquire what was meant. He was too anxious to reach the palace and find his father King Æson, to say nothing of the wicked Pelias.

Jason had left Iolcos when still a small boy, as you will remember, and so he had almost forgotten what the city looked like, but he could still find his way to the palace – which was, in any case, very easy to find, since it towered over all the other buildings. Before he reached it the crowds had thickened, and again he heard the whispers of 'One sandal!' Then the men and women stepped aside, and Jason found himself facing a tall, dark-featured man who did not look in the least friendly. By his side were armed guards, and Jason knew at once that this was King Pelias.

For some moments neither man spoke; each was too busy looking at the other. Then Pelias stepped forward. 'You are a stranger, I can see. Tell me your business, and why you have come to my country.'

Jason drew himself up. Whether Pelias would try to kill him or not, he did not know; whatever might happen, he was not prepared to tell anything but the truth. 'My name is Jason. I am your nephew, and the son of King Æson, the brother whom you have so cruelly wronged. I am here to tell you that you must step down from the throne, and give it back to my father, who should be ruling here.'

Pelias frowned. 'Jason,' he muttered. 'Yes, I remember that my brother Æson had a son of that name, but I had believed him dead. How do I know that you are not lying?'

'It should be easy to prove,' said Jason calmly. 'I have not seen my father since I went to the centaur's cave many years ago, but I have no doubt that he will recognise me as his son. If you are doubtful, take me to him.'

'My brother Æson is some distance away,' said Pelias, and thought quickly to himself: 'This is certainly the one-sandalled man of whom the Oracle spoke.' Aloud, he went on: 'Your noble bearing makes me sure that you are indeed Jason, and I promise you a warm welcome here. Tell me, why have you chosen this time to come to my palace?'

'Because the palace is not yours,' said Jason in a loud, firm voice. 'You are not the king, and you have no right to the throne. I demand that you step down, and restore King Æson to his people.'

Pelias did his best to twist his face into a smile. He was not used to smiling, and all he could manage was a sort of half-snarl, but when he replied his voice was soft and gentle. 'My poor nephew, you are making a great mistake. My brother Æson is alive and unharmed, as you know. He had no wish to rule this country, and it was by his request that I took his place. Tell me,' he went on, without giving Jason a chance to speak, 'what would you yourself do if you were threatened by a man who swore to take away your possessions, your home and even your life? That is the problem facing me at present, and since you have been taught by Chiron the Centaur, who is wiser than any man, you may well be able to give me the answer.'

Jason stared. 'You ask for my advice?'

'Of course,' said Pelias, still in the same gentle tone. 'You are of royal blood, and you have been well schooled. What would you do to such a fellow?'

There was a long silence. Then, at last, Jason said: 'There is only one answer. I should send such a man in quest of the Golden Fleece which lies in the sacred grove at Colchis, guarded by a dragon which never closes its eyes.'

'And if the man refused to go?' asked Pelias, still doing his utmost to smile in a friendly fashion.

Jason laughed. 'If you ordered him to prove his courage, he could hardly say "no" without admitting himself to be a coward. The Golden Fleece may be well watched, but I am sure that it could be taken by a brave adventurer.'

Pelias drew himself up to his full height, and gave a fierce scowl which made Jason look up in alarm. 'Very well,' he said, his voice no longer gentle, but as harsh and cold as the east wind. 'You must know that some years ago, I went to the sacred Oracle and was told that I would be turned off my throne by a stranger with one sandal. You are that man, Jason; you have one bare foot, and you have already told me that I must step down so that your father may take my place. You have given me the answer that I had hoped. Go, then, and bring me back the Golden Fleece!'

Too late, Jason saw the trap into which he had fallen. For a moment he had been tricked by Pelias' soft voice, and now it was impossible to draw back. He did not hesitate; he merely returned Pelias' stare, and said, quietly: 'I accept your challenge, and I will bring you the Golden Fleece. My only condition is that when I return, you will keep your part of the bargain and give up a throne not rightfully yours.'

'That I will promise to do,' said Pelias with a sneer. 'I hardly think that you will come back, with or without the Fleece, but if you can manage to conquer the fire-breathing dragon, to say nothing of the many other dangers that you will meet, you will have earned a kingdom for yourself. Be off with you, and trouble me no more.'

But Jason was not so foolish as to rush away at once. Colchis, where King Æetes ruled, was many hundreds of miles distant, across unknown seas; besides, it would be hopeless to go alone. 'If you mean to keep your word,' he said, 'you will give me two heralds, so that I may gather a band of adventurers to come with me. If you will not, then I shall know that you are mean and treacherous.'

Pelias shrugged his shoulders, and turned back to the palace. 'You can have your heralds with pleasure,' he said as he walked away. 'Collect all the companions you like, provided that they are stupid enough to go with you on this mad search. I will give you woodmen to cut down trees for the ship which I suppose you will try to build, and for all I care you can live in the palace until you set out – it will be the last time that you will set eyes upon Iolcos!'

'We shall see,' was all Jason would reply. 'Remember, the Oracle is never wrong. You have been told that a one-sandalled man will bring you to your ruin, and if I am indeed that man then you cannot escape.'

Inwardly Pelias was badly frightened, but he did his best not to show it; after all, he thought, many heroes had set out in quest of the Fleece, and all had ended up by being eaten by the dragon. In any case, there was nothing he could do except wait to see how Jason would fare.

Jason, for his part, wasted no time. The heralds were sent off within a few hours, and called to all the brave men of Greece: 'Which of you will dare to come in search of the Golden Fleece? Who among you has the courage to risk being scorched by the dragon which guards the sacred grove?' It was not a call which could be ignored, and before long Jason had been joined by all the famous heroes who happened to be anywhere in the country. Hercules was the first; he arrived swinging his mighty club, only too anxious to batter out the brains of any dragon unwise enough to get in his way. Then came Castor and Pollux, the twin boys who were related to Jupiter; Castor was said to be the best wrestler in the whole world, while his brother was famous for his skill at boxing. Next to arrive were Lynceus, whose eyes were so sharp that he could see through the thickest stone walls, and Tiphys, who knew his way about the sky better than most men know their home towns. There was Orpheus, the musician who could charm even the waves with his lyre; there were Mopsus, who knew the language of the birds, and Argus, the shipbuilder. One woman came to join the adventurers – Atalanta, whose skill with bow and arrow was unmatched. And we must not forget Calaïs and Zetes, the two sons of the North Wind, who were particularly welcome, since if asked they could puff out their cheeks and blow a breeze in any direction they chose. There were plenty more heroes, too, and as Jason looked at them he felt proud and confident. As for Pelias, he merely scowled more darkly.

You may imagine that the people of Iolcos were glad to see such a band of brave adventurers – not forgetting Atalanta, of course, who was as fearless as any man. But there was work to be done, and for many days the sound of axe-blows echoed across the plain. When enough trees had been cut, Argus the shipbuilder took charge, and in a remarkably short time a great galley was put together, with positions for fifty oarsmen. All agreed that it should be named the *Argo*, and that Jason should be its captain (Hercules was invited, but refused; as he pointed out, he had more than Jason's strength but less than Jason's wisdom, which under the circumstances was very modest of him).

It was then that Jason had a sudden idea. He knew that some miles away, well out of the city, there grew a very famous tree which was called the Speaking Oak of Dodona. To be quite honest, the Oak had no real voice; but as the wind rustled in its branches it was said that words could be made out, and the tree was sacred to Jupiter, which made it very important. So taking Orpheus the musician with him, Jason made his way to the Speaking Oak, and stood beneath it. Sure enough, the branches began to sway, and after a few minutes the two men could make out a short sentence:

'Cut me off! Cut me off! Cut me off!'

'What does it mean?' whispered Orpheus, hardly daring to raise his voice.

Jason pointed to one particular bough, close above his head. 'Unless I am very much mistaken, we are being told to lop off that branch and place it in the prow of our ship,' he said gravely. 'I am right, wonderful tree, am I not?'

Again the branches waved, and they could make out what seemed to be 'Yes – yes – yes.' So Jason lost no time in cutting off the bough; and as soon as he had done so, the rustling stopped, leaving the Speaking Oak calm again.

It was not long before the Argonauts, as the adventurers had begun to call themselves, had good reason to be glad that Jason had the Speaking Oak. An unexpected difficulty almost made them give up their voyage before it had even started. They had built the *Argo* on dry land, of course, but they had forgotten that it would be very heavy; they could not stir it as much as an inch. All the fifty heroes pulled, and pushed, and tugged, and strained, but without the slightest result. Hercules alone was as strong as ten ordinary men, but at last he gave up, and mopped his brow.

'I know when I'm beaten,' he said in an angry tone. 'We should have thought of this. How we can get the *Argo* launched, I just don't pretend to know. Surely our clever captain must have some idea of how to do it?'

All eyes were turned to Jason, who flushed slightly. This was his first test as a captain, and he only hoped that he had not been deceiving himself. 'Bough of the Speaking Oak,' he said loudly and clearly, 'you can see our trouble. Tell us, please, how we can move our ship into the water.'

For a few moments there was no reply, and Jason was just beginning to despair when he heard the same rustling as he had heard when standing under the tree at Dodona. It was hard to make out what was being said, but at last he thought he could understand, and he called out: 'Orpheus, play upon your lyre while all the rest of us stand back. Give us the most beautiful music that you have ever played!'

'What good will that do?' grumbled one of the Argonauts – I think it must have been either Calaïs or Zetes, the sons of the North Wind, who had very little use for music at the best of times. 'We all know that Orpheus is a splendid player, but tunes won't help us to launch the *Argo*.'

'Wait, and watch,' said Jason quietly. 'Start your song, Orpheus.'

Orpheus did not understand, but was no time to argue, so he took down his lyre and began. There was utter silence except for the sound of his voice and his playing, and even Calaïs and Zetes were spellbound. Then, slowly, the *Argo* stirred. Untouched by any hand, the great ship heaved and swayed, moving towards the sea. Orpheus kept playing, and then, with a splash, the *Argo* was in the water, floating proudly and looking strong enough to carry the crew to the ends of the earth if necessary. A cheer rang out, so loud and joyful that even King Pelias heard it, and came to the edge of the cliff to find out what was going on.

'A plague upon them all,' he muttered, glaring down as though he would have liked to see the *Argo* sink beneath the waves, taking Jason and his companions down too. 'I should have killed this wretched fellow when he first arrived here. Now I am starting to fear that it is too late … Oh well, there's always the dragon. Perhaps Jason will be gobbled up in the same way as all the others who have tried to snatch the Golden Fleece. I very much hope so!'

But neither Jason nor any of the other Argonauts were in the least bothered about King Pelias; already they were at their oars, apart from Lynceus, whose sharp eyes had earned him the post of look-out man, and Tiphys, whose knowledge of the stars meant that he would be busy telling the oarsmen which course to steer. Calaïs and Zetes blew as strong a breeze as they could, and the *Argo* seemed to race off across the sea, moving farther and farther away from Iolcos with every stroke of the oars. King Pelias stood on the cliff and watched them go, the scowl on his face blacker and gloomier than ever before. Even after night had fallen, and the *Argo* had long since passed out of view, Pelias still stood there, wondering how long it would be before he saw Jason again, but still hoping that some disaster would happen to him before he came within striking distance of the Golden Fleece.

Jason had been in such a hurry to get away from Iolcos, before King Pelias could make more mischief, that he had not waited to gather enough food and water for a long voyage. He was not worried about this, because Tiphys had already told him that their way would take them close to Mount Pelion, where Chiron the Centaur lived. Since many of the Argonauts had been Chiron's pupils, it needed no persuasion to make them stop a while in order to see him – and you may well imagine how glad Chiron was; he came clattering down, whisking his tail in delight and promising to provide them with all the stores they needed. He had some new pupils now, and in particular there was Achilles, a small boy who later became one of the world's greatest warriors.

'I wish you could stay longer,' said Chiron, after the adventurers had eaten and drunk, and were stretched out upon the mountain-side warming themselves in the rays of the sun. 'I am never alone, as you know, and I have more than enough work to do, but it is many years since I had so many of my pupils with me at any one moment.' He pawed the ground with his front hoof. 'I hear news of you all the time, of course, and you seem to have learned better than I could have hoped.'

Jason laughed. 'If we are doing good in the world, it is you who should be thanked – tail, hoofs and all!' he said, and there were murmurs of 'Yes!' and 'That's true!' from the others. 'Tell me one thing, if you can. After I left here, I carried an old woman across a stream. Somehow I do not feel that she was the poor old peasant that I first thought, but I cannot be sure. Who was she?'

'I am not certain,' said Chiron, slowly, 'but I think I know. Unless I am wrong, Jason, that old woman was none other than Juno, the wife of Jupiter and queen of Olympus. In any case, I believe that the gods are with you, and that you will be successful in bringing back the Golden Fleece. Do not forget to come and see me on your way back,' he added, a little wistfully. 'If I were younger, and if I did not have so many pupils to teach, I would come with you.'

'Your place is here,' said Castor and Pollux together. 'There are many heroes in Greece, but there is only one Chiron, and we cannot do without your wisdom.'

Chiron smiled. Though he would have liked very much to go along with the expedition, he knew that Castor and Pollux were right, and he said no more, though after the goodbyes were over and the *Argo* had sailed away it took the wise centaur some time to settle back into his usual quiet way of life.

King Pelias had warned Jason that the voyage would be dangerous, but even Pelias could have had no idea of the adventures which the Argonauts would meet before they came anywhere near Colchis. To describe them all would take a very long time, but I really must mention the Harpies, who were hideous birds with skins of iron and the faces of ugly old women. These Harpies were making life a complete misery for King Phineus, who ruled over a small island which happened to be convenient for the spot in which the Argonauts could rest themselves after a long day's rowing.

'I am at my wits' end,' said Phineus sadly. 'No sooner do we prepare food for ourselves than these dreadful creatures swoop down, snapping at us and snatching every morsel of our meal. We cannot hurt them, because their iron skins are too tough, and we are unable to catch them, because they can fly and we cannot. If you can think of any way to get rid of them, I will be deeply grateful.'

Jason smiled, and looked across at Calaïs and Zetes, the sons of the North Wind. 'This seems to be a task for you two,' he said. 'I hardly think that the Harpies will be able to produce as stiff a gale as to blow you away. Are you ready for the fight?'

'Of course we are,' said Zetes, and Calaïs nodded in agreement. 'Truth to tell, things have been rather too quiet for our liking during the last day or two, and personally I feel in the mood to tackle any number of Harpies, iron skins and all. King Phineus, when do you think they will next come?'

'At dinner-time, no doubt,' said Phineus gloomily. 'They always do. They are fierce enemies, I warn you.'

Neither Calaïs nor Zetes minded how fierce the Harpies were, and both looked forward to doing the unhappy Phineus a good turn. Sure enough, the Harpies swooped down as soon as a meal had been prepared, screeching and howling, and scratching with their iron claws. Castor, who was closest to them, swung a club and managed to land a blow on the nearest creature, but all that happened was a dull 'thwack'; for all the good it did, Castor might have been using a feather instead of a heavy stick. Then, with a roar and a bluster, Calaïs and Zetes were in the air, attacking the Harpies furiously.

The battle raged high above the ground, while Phineus and the Argonauts watched in excitement. Strong though the Harpies were, they could not match the speed and skill of the sons of the North Wind, and before long they were flying away as fast as they could, with Calaïs and Zetes in hot pursuit. When the two Argonauts came back, hours later, they explained – rather breathlessly – that they had blown the Harpies so far away that they would certainly never come back, so that Phineus and his people would be troubled no more.

Another danger was that of the Moving Rocks, through which the *Argo* had to pass. These rocks were two in number, and whenever a ship tried to pass between them the rocks would come together, crushing the luckless vessel between them. Fortunately King Phineus knew all about them, and when the *Argo* had sailed away, leaving the islanders to enjoy their meals without fear of interruption from the Harpies, Jason called a meeting.

'Phineus has told me what we must do,' he said. 'Do you see any tall rocks ahead of us?' He pointed. 'My eyes are not keen enough, but I am sure that the rocks must be there.'

'I can make them out very plainly,' said Lynceus, from the ship's prow. 'Indeed, I believe I can see a bird perching upon one of them. There are two rocks, each very tall, and there is shallow water to each side, so that we have no choice but to row between them. They look quite harmless to me.'

'They are anything but harmless,' said Jason grimly. 'Didn't you hear what King Phineus had to say? If we try to sail through, the rocks will come together, and crush us to a pulp.'

There was a short silence, and then one of the Argonauts – it may have been Atalanta – spoke. 'If that is the case, we must try to sail through the shallow water. If we go aground, Orpheus can move the *Argo* back into the deeper sea by playing on his lyre, as he did before.'

'Couldn't he charm the Moving Rocks?' suggested Mopsus, who, as we know, could talk to the birds, and who was also a remarkably wise man.

Jason nodded. 'That is what King Phineus advised. As soon as we reach the rocks, I will release a dove which he gave me. The dove will fly through the gap, and the rocks will try to crush it, but they will not succeed, because the dove flies so fast that it will escape easily. As the rocks start to draw apart, we will row through at full speed, and Orpheus must play his most beautiful music so that the rocks will hesitate before coming together once more. Does anyone disagree?

Nobody did, and by good fortune the whole plan worked excellently. As the dove flew between the two rocks, the great cliffs started to come together, but as soon as the bird had fluttered out of danger the rocks drew back. Then Jason gave a call; Orpheus began his music, and the oars dug deep into the water, sending the *Argo* along at top speed. They were half-way through the gap before the rocks began to close in once more, but as Orpheus' song echoed between the cliffs there was a pause. Now they were nearly through; the rocks had started to move again, and Jason wanted to cry out as he saw the great grey mass so close above him. 'Faster!' he shouted, and the oars flashed. Now they were clear, but only just in time. Behind them the Moving Rocks, unable to stop themselves, had met with a tremendous crash, and had locked themselves so firmly together that they were never able to part again.

The Argonauts paused for breath, and looked back at the danger they had just escaped. 'That was a near thing,' muttered Pollus. 'Another few seconds, and we would have been crushed.'

I really cannot stop to tell you of the other adventures which the Argonauts met on their way, though I must mention Atalanta's feat in chasing and killing a fierce wild boar which had been causing endless trouble to the people of the land of Calydon. Then there were the six-armed giants who constantly attacked the country of King Cyzicus, a friendly ruler who made Jason and his crew very welcome. The giants were of great size, and their six arms made them dreadful enemies, since any one giant could throw stones, launch arrows, swing several clubs, and protect himself with a shield all at the same time. Even so, the Argonauts were more than a match for them, and when the battle was over the giants had been given so severe a lesson that they never bothered King Cyzicus again.

At last, many weeks after setting out, Lynceus called that he could see the coast of Colchis in the distance. 'It's a pleasant-looking land,' he said, 'and I can make out a tall building which I suppose must be the palace of King Æetes. There are many people about, and the city looks very beautiful in the sunlight. Surely you can see it for yourselves?'

Tiphys, the navigator, stared across the water. 'I can see nothing but ocean,' he said. 'Are you sure you aren't imagining things?'

But Lynceus' sharp eyes had not tricked him, and before long the *Argo* had come close enough for Jason to see the land for himself. Even before they reached the water's edge and anchored safely, a crowd had gathered to meet them, and presently there was a fanfare of trumpets as King Æetes himself appeared. Jason stepped ashore, and bowed.

'My greetings to you, King Æetes. My name is Jason, and I am the son of the true king of Iolcos. I come in friendship.'

'Oh, you do, do you?' grunted King Æetes, and looked at the Argonauts sternly. 'I've heard strange things about Iolcos, and I can't imagine why you and your friends have come so far. I doubt it it's merely for the pleasure of meeting me. What do you want?'

Jason returned the king's stare. 'I want the Golden Fleece,' he said bluntly. 'I have promised that I will take it back to my own land of Iolcos. When I do so, my uncle Pelias, who has taken the throne which should belong to my father, has given his solemn promise that he will step down and hand the kingdom back to its true monarch.'

King Æetes scowled, and for a moment he looked just as evil as Pelias himself. 'I never heard of such a thing,' he said angrily. 'The very idea! You come here, unasked and unwanted, and tell me that you plan to take away my most precious possession. Be off, before I set my soldiers on you!'

Calaïs, Zetes and several more of the Argonauts stepped forward, raising their spears, but Jason waved them back. 'Believe me, I am no enemy of yours,' he said in a quiet voice, 'but if you attack us I fear that you will regret it. This ship carries almost all the great heroes of Greece, and if we can deal with an army of six-handed giants, as we did on our way here, we should be more than a match for your soldiers. I would much prefer to come and go in peace, but I am ready to take the Golden Fleece by force if necessary.'

King Æetes glared. For a moment he seemed almost ready to signal his soldiers to attack, but then he paused. 'I think you would find it hard to take the Fleece without my permission,' he said at last, 'but I am ready to say 'yes', provided that you fulfil three conditions. Do you agree?'

'That depends on what the conditions are,' said Jason. He remembered how he had been tricked by Pelias, and he certainly did not want to fall into any more traps. 'If they are reasonable, I will carry out any tasks that you set me.'

King Æetes nodded. 'Very well. First, you must go to the Field of Mars, and plough it up, which means that you must first tame the two fire-breathing bulls made specially for me by Vulcan, the blacksmith of the gods. Then you must plant some of the teeth which once grew in the jaws of a dragon killed by the hero Cadmus long ago.' Æetes gave an unpleasant smile. 'These teeth never fail to produce a remarkably interesting harvest, I assure you.'

'I doubt if fire-breathing bulls can be any worse than birds with iron claws,' said Jason. 'And the third condition?'

Æetes smiled again. 'All that you have to do is go and take the Fleece from its sacred grove,' he said. 'To be sure, there is a dragon coiled round the tree upon which the Fleece hangs, but no doubt a hero as brave as yourself will dispose of such a creature speedily. Well, what is your answer?'

'I agree,' said Jason promptly. 'Tomorrow we will come to the Field of Mars, and I will sow the dragon's teeth.'

King Æetes gave a laugh, and pointed to the north. 'The field lies there, about two miles from the city,' he said. 'I will be there by sunrise, and I look forward to seeing you harness the bulls. They haven't had the chance to scorch anybody for several months now, so I am sure they will be delighted to see you.'

With that, Æetes led his soldiers away towards the palace, leaving Jason standing by the shore staring thoughtfully northward. He had sounded very brave and confident, but inwardly he did not in the least like the idea of facing two bulls which breathed fire and smoke. Then he saw that one of the king's companions had stayed behind, and was walking over to him. Jason drew in his breath sharply. He was facing a young woman, as beautiful as Atalanta but with an expression which was not entirely pleasant.

'Who are you?' he asked.

'My name is Medea. I am King Æetes' daughter, and I overheard all that was said between you.' The girl gave a half-smile. 'It is just as well that I did. You may be brave, Jason of Iolcos, but you are also foolish. If you go into the Field of Mars unprotected, nothing can possibly save you from being scorched to a cinder by the bulls' first breath.'

'If Zetes and I fly above the creatures, and blow their fiery breath towards the ground –' began Calaïs, but Medea stopped him, speaking impatiently.

'That would be of no use at all,' she said. 'Listen to me, if you want to succeed in your quest. I am fond of my father, but he can often be cruel, and I feel inclined to help you. Yet if I do, you must promise to do exactly as you are told, and not to disobey me in anything I say.'

Jason looked puzzled. 'Princess, I am grateful to you, but – well, what can you do? You are a woman; I am a man, and I have faced dangers before.'

Medea turned away. 'Very well. I have no time to waste, so I will leave you to deal with the bulls and the dragons as well as you can. I am sorry that you will never live to go back again to your own country.'

'Wait,' broke in Lynceus, who had been staring across the plain. In the distance, so far away that nobody else could possibly see it, he had been able to make out a grassy field in which were two large, fierce-looking animals. 'Tell me, Princess Medea, are you an enchantress?'

Medea paused, and then broke into a deep laugh. 'You are no fool, I can see. You are right, of course; I am an enchantress, and if I protect you with my spells you will have at least a chance of gaining the Golden Fleece. Still, if you prefer to go your own way, I will not try to stop you.'

Not for a moment did Jason doubt that Medea was telling the truth. Something about her showed clearly that she was no ordinary princess, and although he did not quite trust her he could see that she would be a very powerful friend. So he said, humbly: 'I am sorry. Tell me, please, what is to be done? I promise that I will do as you say.'

Medea looked round to make sure that the soldiers and the people of Colchis were out of sight, and then she drew out a bottle. 'Take this,' she said. 'Tomorrow morning, before you start out, smear yourself all over with the ointment. I knew that you were on your way here – do not ask me how; it is not important – and so I prepared the ointment specially. It will protect you from anything that the bulls can do, but you must cover yourself completely. Do you understand?'

Jason took the bottle, and nodded. 'If the fiery breath cannot hurt me, I am quite sure that I can tame the bulls and plough the Field of Mars, but what about these teeth that your royal father spoke about?'

Medea smiled, looking no more good-natured than King Æetes. 'The dragon's teeth are strange things,' she said. 'The dragon itself was killed by a hero named Cadmus, who built this city and became its first king. When planted, the teeth produce armed soldiers, who are very apt to turn upon anybody within reach and cut him to pieces with their swords. I will tell you how to deal with them when the time comes. Well, do you still mean to try to obtain the Golden Fleece? If you are in the least afraid, I advise you to get back to your ship and make off as fast as your oars can row you.'

But Jason had no thought of turning back, and it seemed that even if Medea were ill-natured (as most enchantresses are), she could hardly be plotting against him. So next morning he took a long time in covering every every part of his body with the magic ointment, after which he put on his armour and made his way to the Field of Mars, where King Æetes was waiting for him together with his soldiers and a great many people who had come along to watch. Medea was there too, and Jason gave her a quick nod.

'So you have come after all,' said Æetes. 'Well, well! I admit that you have more courage than I would have expected. Let loose the bulls.'

The captain of the soldiers shouted an order, and then, with a snorting and a roaring, two huge animals rushed on to the field. They were much larger than ordinary bulls, and instead of breathing in the usual way they sent out clouds of smoke, together with flames which scorched the grass and turned it black. Even Jason felt unwilling to go anywhere near them, but then he saw a mocking smile on Medea's face, and he hesitated no longer. 'If I'm going to be burned up, I may as well get it over quickly,' he thought, and after bowing to King Æetes he marched firmly off into the field.

The nearest bull gave an ear-splitting bellow, and rushed forward. Jason was surrounded by red flame, and the smoke made him cough and choke, but he could not feel even a slight warmth; Medea's ointment protected him from any harm, as she had said it would. Grasping the bull by the neck, Jason forced its head downward. Then the second bull was upon him; Jason gripped it with his other hand, and somehow he seemed to have as much strength as Hercules himself. Though the bulls kicked and struggled, they could not break free. After some minutes they quietened down, and the flames from their nostrils died away until they stood there quivering and looking thoroughly beaten.

'Upon my word!' muttered King Æetes. 'This man Jason must have had help from some witch or other. Nobody has ever vanquished the bulls of Vulcan before.'

He looked hard at Medea, but the princess made no reply, and merely watched as Jason fetched the plough, harnessed the now-obedient bulls to it, and started to plough the field into neat furrows. It took him an hour or so, by which time King Æetes had more or less recovered his good humour. Jason might have tamed the bulls, he thought, but it would take more than a magic spell to save him from the soldiers of the dragon's teeth.

At last Jason finished his work and released the bulls, which trotted away looking very subdued – and also, to be truthful, much happier than when they had been breathing fire; after all, their noses must always have been uncomfortably hot. King Æetes scowled, and handed over a large leather bag. 'Here are the teeth,' he said. 'Plant them, and watch them sprout! Soldiers, make ready.'

The soldiers pulled out their spears, and arranged themselves in formation all round the Field of Mars, so that anyone inside would be unable to get out. Jason stepped forward, and walked to and fro, scattering the huge teeth in the furrows; each time he dropped a tooth it seemed to sink into the earth, and disappear without any plough having to be drawn over it. There were thirty or forty teeth in all, and he took good care to sow each one, after which he stood back and waited to see what would happen.

There was a long pause. King Æetes said nothing, and neither did his people; even Calaïs and Zetes, who had been watching the scene from overhead (against Jason's orders, it must be admitted; he had told the Argonauts to stay behind on board ship), were silent. At last Jason fancied that he could see a silver glint in the middle of the field, and he gripped his sword. Another glint followed, and another, and another; the brilliant specks grew in size, and before long a whole host of helmets had appeared. Next came human faces, and within minutes a whole troop of armed men stood on the Field of Mars, shaking their feet free of the earth and shouting to each other, clashing their swords and uttering loud battle-cries.

'Who has called us up?' they roared. 'Where is he? Lead us to him, so that we may run him through. We are ready to fight any god or man, so beware of us!'

They went on shouting and calling, and then the nearest warrior – a tall, broad fellow with an ugly expression and gleaming eyes – caught sight of Jason, who was the only other man on the Field of Mars. 'There he is. Attack, my brothers! Fight, fight, fight!'

The whole mass of men turned, and began to rush towards Jason, who raised his own sword and wondered whether he would be able to kill more than a few of these quarrelsome beings before he was overwhelmed. Then he heard a voice in his ear, and realised that Medea was by his side. 'Here is a stone,' said the enchantress softly. 'Throw it into the middle of the crowd. Quickly – it's your only chance!'

Jason did as he was told, and the stone sailed through the air, landing on the head of one of the tallest warriors. The effect was surprising. The onward rush stopped, and a voice rang out: 'Who hit me? Come and fight, whoever you may be, or else I will know you to be a coward!'

Medea threw another stone, and again there came a loud 'clang' as it struck a helmeted warrior. In a second the fight had begun, and all the men of the dragon's teeth were cutting, stabbing and clawing at each other, hacking off arms, legs and even heads. The noise was so tremendous that King Æetes, watching from a safe distance, shivered uneasily, while Jason felt deafened. Only Medea, standing close beside him, seemed to be really enjoying herself.

The fight raged on and on, but there were fewer men left standing now, and at last there were only two; with a final clash of arms they ran each other through, spun round, and fell to the ground, giving a last roar before tumbling down in lifeless heaps. Silence fell; the Field of Mars was littered with dead bodies, and not one of the warriors of the dragon's teeth remained alive. Their whole career on earth had lasted for less than half an hour.

King Æetes climbed down from his chariot, and walked across to Jason, his face as black as thunder. 'You have had magic help,' he said harshly. 'Tell me, who is the cause of all this? No man since Cadmus has ever sown the dragon's teeth and lived to tell the tale.'

'That is my own affair,' said Jason; Medea had moved away, and he was not in the least inclined to tell Æetes who had protected him. 'Well, your majesty, I have carried out my first two tasks. Are you now ready to keep your part of the bargain and let me go to the grove where the Golden Fleece hangs?'

'I – I suppose so,' muttered King Æetes. 'Very well. Nothing more can be done today, because I have a great many things to attend to, but tomorrow morning I will come with you to the grove. Now be off back to your ship, before I change my mind.'

In some ways Jason would have liked to have gone straight on to the sacred grove, but he did not want to make Æetes any angrier than was necessary – and besides, he wanted to have another talk with Medea. The magic ointment had served him well against the bulls, but it would hardly protect him against a dragon, and since the enchantress had already helped him once she would presumably do so again.

So Jason merely bowed, and made his way back to the *Argo*. Calaïs and Zetes had seen all that had gone on, so that there was no need for Jason to tell the story all over again. Instead, he ordered his companions to make the ship ready for a quick departure. Should he succeed in overcoming the dragon and taking the Golden Fleece, he thought it very likely that King Æetes would break his word and try to keep him back by force.

The day passed quickly; there was much to be done, and the Argonauts were busy all the time. There was no sign of Medea, but now and then Lynceus, sitting on one of the mastheads, reported that he could see hasty preparations going on in the city; King Æetes was inspecting his soldiers, gathering men from all parts of his country, and evidently making ready for an attack of some kind or other. 'He seems to be in a bad temper,' said Lynceus. 'I think there is no doubt that he means us to be his victims.'

'We'll see about that,' growled Castor and Pollux. 'If he sends his men against us, he'll be given a warm welcome!'

Night came at last, and the stars shone down; Jason could see the pattern of the faithful ram which had carried Phryxus to Colchis, and whose fleece now hung in the sacred grove. Soon after dark he heard a call, and he replied softly: 'Who is there? Is it the Princess Medea?'

'You knew that I would come,' said Medea, and Jason found her standing beside him, though he did not know how she had crossed the water. 'I have bad news for you, Jason. My father has been so upset by your taming of the bulls and your treatment of the soldiers of the dragon's teeth that he has decided not to let you go anywhere near the sacred grove. In a few hours' time his soldiers will swarm on board and do their best to put all of you to the sword.'

'Let him try!' said Atalanta, and fingered her spear.

'Don't be foolish,' said Medea impatiently. 'You might well win the fight, but you would lose many of your crew, and you would certainly not lay hands on the Golden Fleece. I am ready to help you again, but there is no time to lose. Make your ship ready for departure, and bring it close in to the shore, so that you can row off at a moment's notice. Jason and I will go to the sacred grove, together with Orpheus. Are you ready?'

'I am ready,' said Jason at once, 'and I knew that Orpheus will not fail me. We are under your orders, Princess. Am I to smear myself with more of your ointment?'

Medea shook her head. 'No ointment, even mine, could protect you from the dragon which guards the Golden Fleece. We must try another trick this time; I will explain when we reach the grove. Come, then, and be as quiet as you can. If the soldiers hear us, you will not escape.'

Jason and Orpheus did not argue, and followed Medea ashore, leaving the rest of the Argonauts to make the ship ready. Everything was inky-black, and even Lynceus would have had some difficulty in finding his way about, but Medea was in no trouble; like most enchantresses she could see in the dark, and was as happy in utter blackness as in the glare of the sun. Quietly she led Jason and Orpheus along, past the royal palace, past the city and past the Field of Mars. As they crossed the field, Jason felt a soft muzzle in his hand; it was that of one of the bulls, now grazing happily on the grass outside instead of scorching it.

'Their fires have gone out,' said Medea with a chuckle. 'My father is not pleased, as you may well imagine. Watching people being burned up was one of his favourite hobbies. I admit that I rather enjoyed it myself.'

Jason had the sense not to reply, and the three passed beyond the great field, entering a thick wood. No insects hummed among the trees, and altogether the wood seemed to be a very strange place indeed. Ahead they could see a faint glow, and suddenly they came to a clearing. In the centre was a large tree, and from the tree hung a shining mass which could only be the Golden Fleece.

Jason paused. 'It's beautiful,' he said softly. 'Why should I not simply go and get it?'

'Look beneath,' said Medea, and pointed. 'One step more, and you will be gobbled up.'

Jason stared in amazement. He could see that the enchantress was right; at the foot of the tree there lay a huge dragon, with a wide mouth filled with snapping teeth. It, too, breathed out fire, and its black eyes were fixed upon the three figures at the edge of the glade. Slowly it reared up, and Jason felt for his sword. 'I can kill it,' he muttered.

Medea pulled him back. 'Even if you ran it through with your sword a hundred times, you still could not harm it. What my father did not tell you is that the dragon is immortal, and cannot be hurt by any man – even a hero such as yourself,' she added in a tone which was by no means pleasant. 'Try, if you doubt me, but I advise you against it. Orpheus! Play your lyre – not loudly, or the soldiers will be on us, but enough to make the dragon hear.'

Orpheus plucked the strings, and a low, soft melody floated over the glade. At first the dragon took no notice; then it stretched out, uncoiling its huge body, and began to sway to and fro with the music. Its mouth opened, and Jason felt Medea thrust something

Jason snatches the Golden Fleece

into his hand. 'Next time you have the chance, throw this into the dragon's mouth. Be careful – I have had no time to make up more than one spell, and if you miss your aim there will be no second chance.'

Jason set his lips, and waited. Still the dragon swayed, and again its mouth opened, showing a row of the wickedest teeth that you could possibly imagine. Then Jason held his breath, and threw. His aim was true; the dragon's mouth snapped shut, and the hideous creature tumbled down, sprawled in huge coils around the tree.

'Quickly,' said Medea, in an urgent voice. 'It's only a sleeping mixture, and it will wear off in a few moments, so you've no time to lose. Take the fleece, and run back to your ship without wasting a second. At any moment now my father's soldiers may start to attack!'

Jason did not wait. He bounded across the glade, kicking against the dragon's coils, and tore down the Golden Fleece, throwing it across his shoulders. Then he turned, and began to race back in the direction of the *Argo*, Medea and Orpheus by his side. There was the Field of Mars; there was the city, with its royal palace – and there was the *Argo*, ready for him by the sea-shore with the fifty oarsmen in their places. The instant that Jason leaped on board, the oars dug water; there was a splashing and a rippling, and the *Argo* drew clear. At the same moment there came a fierce howl, and the first of King Æetes' soldiers appeared on the beach, waving their swords and calling out, 'Come back! Come back, by order of our royal master!'

Jason stood on the deck, and waved. 'Tell your master that I have no further time to waste on him,' he shouted. 'I have kept my word, even if he did not mean to keep his. The Golden Fleece is ours, and you will see it no more!'

There is really very little to add. Of course the Argonauts had many more adventures on the way home, but they arrived at last, and Pelias, little as he liked the idea, had no choice but to step down from his throne and hand the country back to King Æson. By that time Æson was very old and ill, but Medea soon changed all that; she was a powerful enchantress, and she cast a spell which made the king young and strong again. Æson reigned for many years, and Jason reigned after him. As for the Argonauts, they went their separate ways, but they never forgot that wonderful voyage when they had joined Jason in quest of the Golden Fleece.

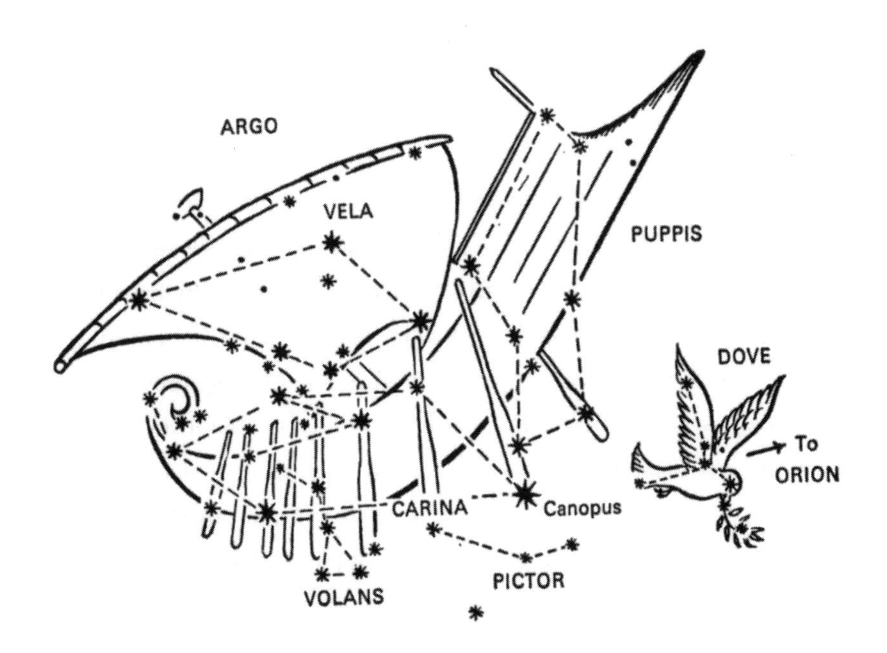

It is a pity that no constellation is named after Jason himself, but at least the great ship *Argo* is there. Only part of it can be seen from Europe; you will find it not far from Orion, but its brightest stars lie far south and so never rise over England. Canopus, the most brilliant of all, is a lovely star which is thought to be 10,000 times as luminous as our Sun, and to be an immense distance away.

The *Argo* was a very large ship, and the constellation named after it is also very big – in fact, it covers so much of the southern sky that astronomers have now divided it up into several parts: Vela (the Sails), Puppis (the Poop) and Carina (the Keel). Carina contains the brightests stars, including Canopus, but all that can be seen from England is a part of Puppis, together with one star, Suhail Hadar, which is about as bright as Polaris in the Little Bear.

Of the heroes, Hercules has already been described. Castor and Pollux, the Heavenly Twins, lie near Orion. The constellation which contains them is named Gemini (the Twins), but Castor and Pollux have each been given a bright star, and you will be able to recognise the pair without difficulty.

Of the two, Pollux is slightly the brighter, but it is thought that Castor may have faded slightly during the past 2,000 years. This, too, has a legend; it is said that Pollux was immortal, while Castor was not. When Castor was killed, Pollux appealed to Jupiter and was allowed to share his immortality with his brother, after which both were placed in the sky.

The Ram (Aries) is there, of course, close to the square of stars which makes up the Flying Horse, Pegasus. You will find them both in the star-map on pages 48 and 49.

And lastly we must not forget the dove which Jason sent through the Moving Rocks. Astronomers call it by its Latin name of Columba, and you will find it below the feet of Orion.

Of all the old stories, that of the Argonauts is perhaps the most fascinating of all. I have been able to tell you only part of it, but I hope I have said enough to give you at least some picture

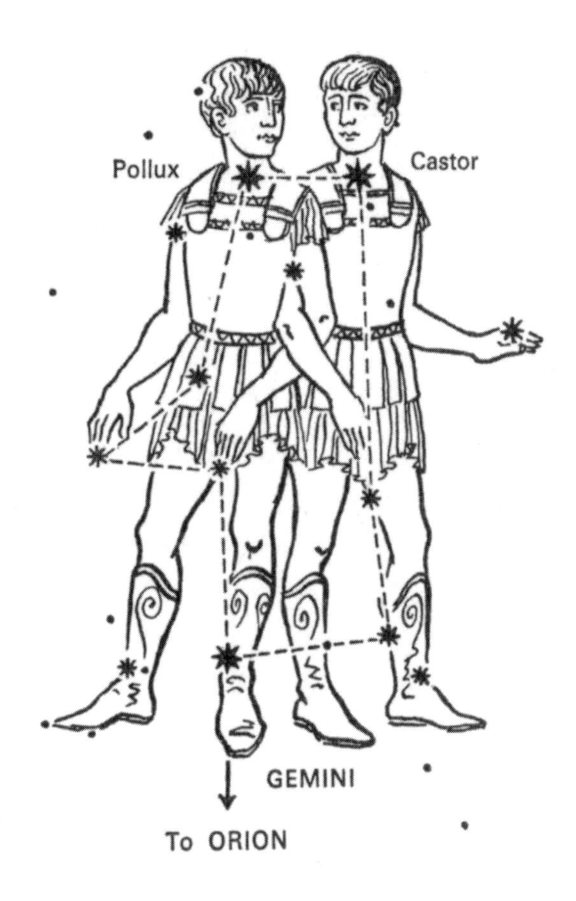

of those gallant heroes who came together in order to undertake a voyage more wonderful than any which we can ever make for ourselves.

I hope, too, that you have enjoyed reading about the gods and heroes of Greece. We no longer believe in the Olympians themselves, as the Greeks did, but at least we can read about their adventures. The stories have lasted 2,000 years and more; we may be sure that they will never be forgotten.

Index

INDEX

INDEX

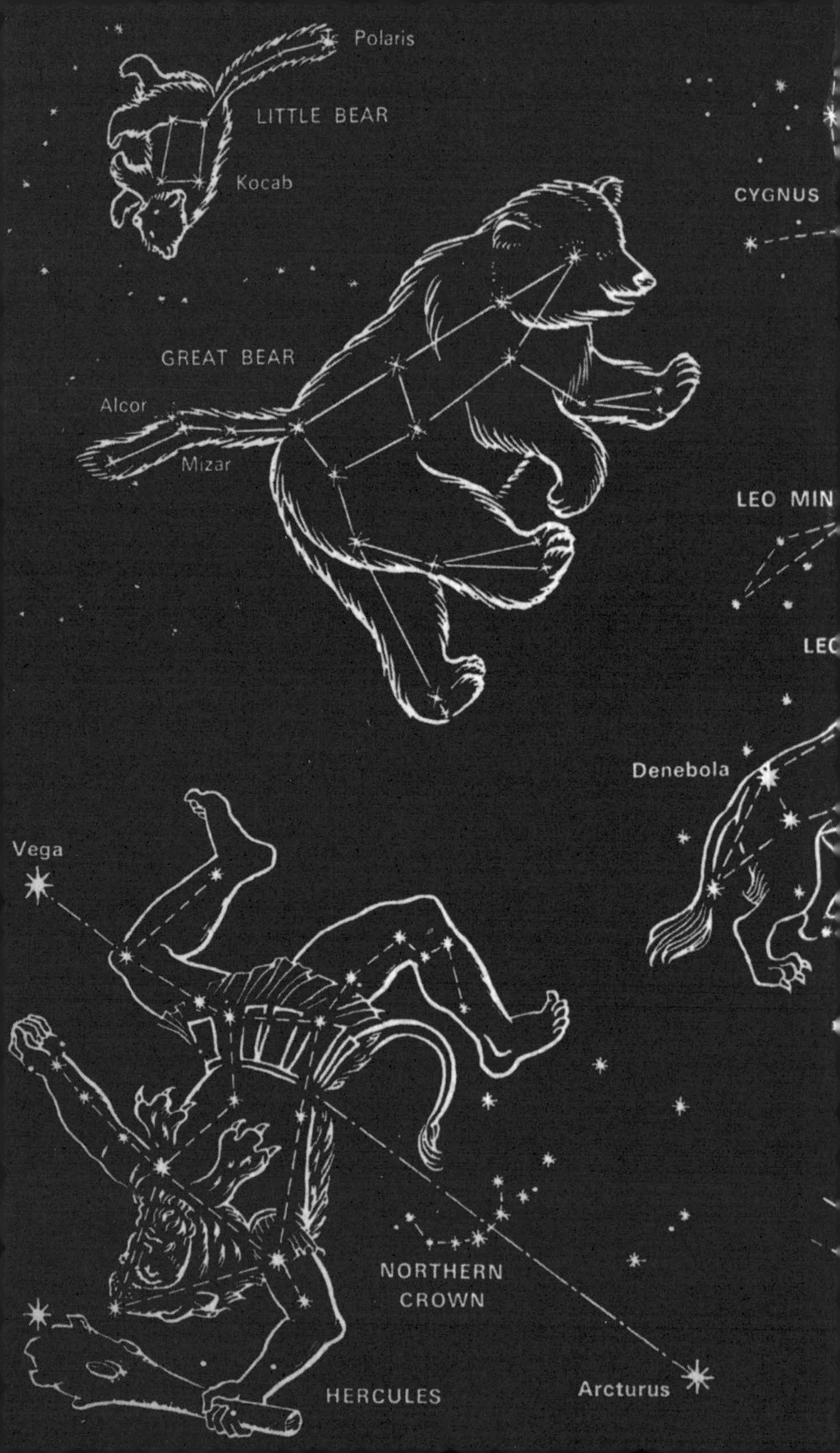
Polaris
LITTLE BEAR
Kocab
CYGNUS
GREAT BEAR
Alcor
Mizar
LEO MIN
LEO
Denebola
Vega
NORTHERN
CROWN
HERCULES
Arcturus

Markab
Enif
PEGASUS
Alpheratz
SERPENS
OPHIUCHUS
LYRE